MACHINES WITH A PURPOSE

MACHINES WITH A PURPOSE

HOWARD ROSENBROCK

Oxford New York Tokyo
OXFORD UNIVERSITY PRESS
1990

Oxford University Press, Walton Street, Oxford OX2 6DP

Oxford New York Toronto
Delhi Bombay Calcutta Madras Karachi
Petaling Jaya Singapore Hong Kong Tokyo
Nairobi Dar es Salaam Cape Town
Melbourne Auckland

and associated companies in
Berlin Ibadan

Oxford is a trade mark of Oxford University Press

Published in the United States
by Oxford University Press, New York

British Library Cataloguing in Publication Data
Rosenbrock, Howard
Machines with a purpose.
1. Philosophy related to science 2. Science related to philosophy
I. Title
100

ISBN 0-19-856346-9

Library of Congress Cataloging in Publication Data
Rosenbrock, H. H.
Machines with a purpose / Howard Rosenbrock.
Includes bibliographical references and index.
1. Technology—Philosophy. 2. Man-machine systems. 3. Control theory. I. Title.
T14.R599 1990 601—dc20 90-36423
ISBN 0-19-856346-9 (hardback)

Typeset by Latimer Trend & Company Ltd

Printed in Great Britain by
Courier International Ltd
Tiptree, Essex

Preface

Dr Johnson maintained that

There are two things that I am confident I can do very well: one is an introduction to any literary work, stating what it is to contain, and how it should be executed in the most perfect manner; the other is a conclusion showing from various causes why the execution has not been equal to what the author promised to himself and to the publick.

It does not need Dr Johnson's skill to describe the aims of this book; and the extent to which it falls short will no doubt be readily apparent. The aims are not modest, and so complete success is not to be hoped for.

There is at present a widespread unease about the direction in which our technology is taking us, apparently against our will. We are damaging our environment, polluting the sea and the air and the land. And in the production of goods and services, we are reducing human work to the trivial mechanical repetition of actions which have no human meaning. If there is much that still escapes from this condemnation, there is nothing to suggest that progress towards it will be reversed.

An attempt to develop the beginnings of a better kind of production technology has been described elsewhere.* It encountered some technical problems, and some problems in crossing interdisciplinary boundaries, but neither of these seemed likely to be a very serious impediment to achieving its aims. What was much more difficult was to see how to give an intellectual defence of what was being attempted, adequate to persuade any large numbers to support it and to prevent it from being subverted by the orthodox view.

The terms in which we have to analyse and develop our technology are those of science, and it came to seem that within the framework which this provides no intellectual defence could be given. The terms in which we think imply and entail the direction in which we move. Nor do the current moral or religious systems offer any better support.

Some have reacted to this perception by proposing to reject our science and technology: but that would return us to a past in which the majority must suffer under a poverty and an intensity of unremitting labour which are as discouraging as what we have. The question then is, can we reject those aspects of our scientific and technological culture which lead to the damaging and inhuman consequences which we wish to avoid, while still

* H. H. Rosenbrock (ed.), *Designing human-centred technology*, 1989 (Springer-Verlag).

retaining enough to escape the deprivations of the past? It is the aim of this book to show that we can.

It is not to be expected that so ambitious an aim will be completely fulfilled, but this provides no justification for avoiding the attempt. The argument is presented in terms which should be intelligible to a wide audience. Technical support is necessary to satisfy the specialist, and this is relegated to two Appendices.

The views put forward are likely to seem contentious, and even provocative. This is by necessity rather than design, and if some reject it on that account, others with a more open mind may be stimulated to follow this or other lines of thought leading in a similar direction.

Ross-on-Wye HR
April 1989

Contents

Appendix 2

Brief development of the stochastic variational treatment of quantum mechanics

1 An outline of the argument

This chapter gives a brief outline of what will be discussed in the book, so that the reader can see how later developments contribute to the argument.

1. *Purpose and causality*

'In science, man is a machine; or if he is not, then he is nothing at all.'[1] This was written by a distinguished biochemist and sinologist with wide human sympathies. It defines his view of an imperative imposed by science, rather than uniquely characterizing his own beliefs. It was also written in opposition to views which can easily be seen as obscurantist. Yet it contains within it the seed of a highly damaging intellectual development which will be our main concern. It is therefore a convenient starting point for our discussion.

What first of all, is meant here by the statement that 'man is a machine'? It asserts that all human behaviour, conscious and unconscious, can be explained in causal terms, 'causal' being interpreted in one of the four meanings distinguished by Aristotle.[2]

i. The 'material cause' of a thing is the substance out of which it is made—the marble from which the sculptor creates a statue.

ii. The 'formal cause' is the essence of the thing itself—what the statue represents.

iii. The 'efficient cause' is what produces a change, an effect—the action of the chisel on the marble under the influence of the sculptor's mallet.

iv. The 'final cause' is the purpose to be attained—the end which the sculptor has in view as he works on the marble.

The first and second of these would not nowadays be called causes, and the fourth is usually distinguished as a 'teleological explanation', or an explanation in terms of purpose. Only the third would now be termed a cause, and for more emphasis an explanation in terms of efficient causes is called an explanation in terms of 'cause and effect'. In specific disciplines, other special terms are used: in biology or psychology 'stimulus–response behaviour', and in engineering 'an input–output description'. Both terms imply a hidden, more detailed, causal mechanism which produces the response or output from the stimulus or input.

What Needham is asserting is that all biological systems, and all human behaviour, can be explained in terms of cause and effect. A given cause produces a certain effect, and this again acts as a cause to other effects, and so on in an infinite causal chain. The claim is not that we can actually describe every link in such a chain, nor even that we shall some day be able to do so. It is only that in principle such an explanation can be given, and no other is needed: that in fact causal relations express accurately and completely the way the world works, and not only for machines but for people.

The vitalists, against whom Needham wrote, maintained on the contrary that causal explanations were not enough in biology. Machines could be explained entirely by cause and effect. Biological systems required in addition, for their explanation, a reference to purpose. Their behaviour could not, even in principle, be fully explained in causal terms: 'Even in the most elementary and general manifestations of life we have to do with purposive phenomena, quite distinct from all the phenomena of the inorganic world.'[3]

This is a debate which goes back to classical antiquity, and Needham was expressing the view that has predominated during the whole period of modern science, which we can date somewhat arbitrarily from 1600. In this view, purpose does not exist except as an appearance, an epiphenomenon. We seem to perceive purpose in ourselves, in other creatures, or in the progress of evolution. But when we study these scientifically, we find that the idea of purpose is redundant. The phenomena which seem to indicate purpose can all be explained causally.

Our obstinate human belief that purpose is not an illusion arises in this view from our failure to be true scientists:

> Its teleology appears to them [sc. working biologists] as a faint thread . . . and only present in them so far as they are imperfect scientists, because they cannot altogether divest themselves of the common and vulgar ways of thought.[4]

> The former [sc. teleology] is only present in the mind of the scientific worker because he is still to some extent an ordinary man.[5]

The rejection of purpose in these quotations goes beyond the statement that it is not needed, and becomes a recommendation which we are urged to follow. Teleological explanations are the mark of the common man. Though we all have something of the common man in us, as scientists we must strive to rise above his way of thinking. But if we take Needham at his word, he himself is a machine, explainable in causal terms like ourselves. Between two such machines, what is the force of a recommendation to follow one course rather than another? To do this is to have a purpose, and purpose has already been excluded.

2. *Control theory*

In a discussion which has been carried on over millennia rather than centuries, can anything new be said? The reason for believing that it can is that during the past forty years a highly developed theory of automatic control has arisen, which shows how to incorporate human purposes in machines. In the light of this theory some new things can be said about the problem, and some of the things which have been said about it in the past can be seen to be false. The theory is described in a later chapter, but some of its leading ideas are described here to show their bearing on the discussion of purpose.

One of the leading ideas is the distinction between a 'schedule' and a 'policy'. Let us invent a simple control problem. We wish to sail a boat from a given starting point to a given finishing point in a given time, using the least amount of fuel. We know how the speed through the water, and the fuel consumption, vary with engine speed. In still water the obvious course, since fuel consumption increases more than proportionally with speed, is to sail a straight course at the constant engine speed which accomplishes the journey in the given time. A straight course is the shortest, while a varying speed would incur more consumption when it was above average than would be saved when it was below average.

If the tide sets in different directions and at different rates at different points, a straight course and constant engine speed will no longer be appropriate. We can solve the problem in principle by specifying an arbitrary course and some variation of engine speed along it (taking care that this accomplishes the journey in the specified period), and working out the fuel used. Then by comparing many such calculations we can select the solution which gives the least fuel consumption. Better, we can use the mathematical techniques of the 'calculus of variations', which have existed for three centuries,[6] or the more recent techniques of control theory described in Chapter 2. By using a digital computer, the course, and the engine speed at each time along it, can be calculated before we start. Then it is only necessary to devise a controller which keeps the heading and the engine speed at the appropriate calculated values at each instant.

We have assumed so far that there are no random disturbances. In fact, the tide will not run exactly as we predicted, and there can be other disturbances from wind, from variations in engine performance, and so on. We shall not be able to predict these accurately, but we may have some information about their probability.

With this kind of probabilistic information, we cannot predict the actual amount of fuel which will be used on any occasion, but we can calculate the average amount which will be used on a large number of trips under the given conditions. We can then look for two different types of control, depending on the measurements which we are allowed to take.

i. A *schedule* gives the best course and engine speed at each instant during the journey. It can be calculated before we start, just as above, and we adhere to it regardless of what actually happens. The schedule is 'best' in the sense that, under the given conditions, it gives the lowest fuel consumption on average, taken over a large number of journeys, which can be obtained when we have no information about the results of our actions during the course of the journey.

ii. A *policy* takes account of what actually happens. It requires information about position, which we can suppose to be measured continually by means of radio transmissions. We then ask the question, 'If at this time I am at the point just measured, what is the best course and engine speed to adopt?' Here 'best' again means that on average over many trips we shall use the least amount of fuel.*

Even when disturbances are present, a policy is deterministic: at each position and time the heading and speed are fixed, because we have averaged out the effects of disturbances. A policy is often called a 'closed-loop control', because there is a loop from measured position to resulting action to consequent position. A schedule, in contrast, is called an 'open-loop control'. Policies use more information than schedules and generally give better results. They cannot give worse results, since we always have the option of rejecting the extra information, which gives us back the schedule. A complete range of intermediate controls can be generated by measuring position only at certain distinct times during the journey. As the number of times we measure position increases from zero, we progress from the schedule towards the policy described above. When there are no random disturbances the schedule and the policy give exactly the same result: we learn nothing by measuring the position which we cannot compute exactly by dead-reckoning.

Now the reason for the foregoing discussion is as follows. A schedule specifies all the future actions which are needed to fulfil the purpose, even when random disturbances are present. It therefore seems at first sight that it requires knowledge of the future. Similarly, the actions prescribed by a policy are those which produce a desired result in the future, and so they again seem to require foreknowledge of the result.

* The definitions of schedule and policy raise some technical problems. These do not affect the discussion here, and they are therefore postponed to Chapter 5.

The easiest way to see that knowledge of the future is not required is to consider the policy. This tells us how to act—what heading to sail and what engine speed to use—for every position in which we can find ourselves and at each time. We therefore need only information about the present situation in order to follow the policy, and this was calculated using only probabilistic information about the disturbances. Knowledge of future events is not needed at any point. The question then arises, what can be said about the cause of our behaviour when we follow the policy?

3. *What can we assert?*

Someone observing our actions from outside, and knowing nothing of the reason for them, could with sufficient effort deduce the relationship which we are using between position and time, as the causal factors, and heading and engine speed, as the effects. He would thus recover the causal relationship which we deduced from the original purpose. This causal relationship makes no reference to the future but relates present actions to present causal factors. (More generally, the causal factors can be in the past, which is the case for a schedule.)

Would the external observer be justified in arguing that because our actions can be explained causally, they do not express any purpose? Clearly not, because here the purpose preceded the causal relations, which were deduced from it. But suppose that he did not know we existed, and yet by further effort had discovered that his causal laws, obtained by observation, could be deduced from a purpose. We can ask what he would then be justified in asserting.

This is very much the situation in large areas of physics. If we throw a stone, its motion can be explained causally by Newton's laws. For a long time it has been known that the motion can also be deduced from a purpose (Hamilton's principle[7]) which has a resemblance to the one we used above. Our boat moved from an initial point to its final point during a given time, and in such a way that (in the absence of disturbances) it minimized the amount of fuel used for the individual journey. From this purpose we could deduce the policy which would fulfil it. The stone similarly moves from its initial position to its final position, in some fixed time, and in doing so it follows the path which minimizes* something called the 'action', which is analogous to the fuel consumption of the boat. The policy which can be deduced from this purpose is expressed by Newton's laws of motion. We are therefore, in observing the stone, very much in the same situation as our imagined observer studying the boat. What are we justified in asserting in this situation?

* Again there are technical questions which are not essential here, and they are therefore postponed to Chapter 3.

The position we shall adopt is that we are not justified in asserting that the purpose pre-existed in some mind. Our explanations of nature, whether causal or purposive, are human constructs. We can explain the motion of the stone by means of Newton's equations, but we do not suppose that the stone solves these equations in order to determine its path, nor that any intelligence informing the world solves the equations on the stone's behalf. The stone goes where it does, and it is we who 'explain' its motion. In the same way, if we explain the stone's motion as fulfilling a purpose, the explanation is ours.

On the other hand, we are not justified in saying that causal relations such as Newton's laws express the true nature of the world, while from explanations in terms of purpose (Hamilton's principle) we must never draw any conclusions about the nature of reality. The two explanations are scientifically equivalent: all evidence which supports one supports the other, all evidence which contradicts one contradicts the other. If we claim that one describes the world in a way which the other does not, we are claiming an a priori knowledge which goes beyond any presently conceivable evidence.

The three questions touched upon here will underlie most of what follows. In brief, the responses which will be given are these:

i. It is often asserted that explanations in terms of purpose (for example Hamilton's principle) imply knowledge of the future. This is simply wrong, as we can show from control theory and related modern developments. The purpose is to be fulfilled without any knowledge of the future, and the mathematical methods which we use, in translating purpose into the actions which will fulfil it, ensure that this is so.

ii. Purposive explanations of nature have been used to support theological propositions. We reject these as going beyond the evidence.

iii. For the same reason we reject the claim that causal explanations contain some essential truth which explanations in terms of purpose do not.

There are interrelations between the questions: for example, if purpose required knowledge of the future, this would have a bearing on the theological argument.

4. *How we behave*

The way we explain our behaviour determines the way we behave. This is why the questions raised above have a more than intellectual interest.

If we explain the world in causal terms, which we believe to express the true nature of reality, and if we reject explanations in terms of purpose as mere scientific curiosities, certain things follow. Our own inner perception of purpose becomes an illusion, an epiphenomenon arising from a strictly causal operation of our minds. Purpose in other people equally becomes illusory, and all moral judgements which refer to purpose are invalid:

> The scientist, *as such*, has no ethical, religious, political, literary, philosophical, moral or marital preferences. That he has these preferences as a citizen makes it all the more important that he dispense with them as a scientist. As a scientist he is interested not in what is right or wrong or good or evil, but only in what is true or false.[8]

> For the scientist there is only 'being', but no wishing, no valuing, no good, no evil; no goal.'[9]

These propositions are stated for 'the scientist', as though they represented an attitude which could be put on when entering the laboratory and put off again when leaving. Few of us have this ability, and those who never enter the laboratory may still be at risk of adopting the outlook of 'the scientist' and carrying it into their daily life.

It is much easier for us to convince ourselves that purpose is an illusion in inanimate things, or in other people, than in ourselves. Then we are apt to see our own purpose, our own will, as unique, operating in and upon a world which responds to our manipulation. Strong influences of this outlook will be traced below in the development of our present technology.

The outlook has strong connections with the idea of ownership. A ruler owns his people if they have no purpose opposed to his will. A slave is owned when he expresses only the purpose of his master. An employer fully owns the time which he has bought from a worker when he can direct in detail everything that is done:

> Under our system [F. W. Taylor's 'Scientific Management'] the workman is told minutely just what he is to do and how he is to do it; and any improvement which he makes upon the orders given to him is fatal to success.[10]

Here the image of the worker as a causal device, a purposeless machine, operating under the control of management, is very clear, and it is no accident that Taylor called his system 'Scientific Management'. The congruence between this outlook, and the picture of a purposeless, causal world is so great that those brought up in the scientific culture find it difficult to question. Engineers, or computer scientists, engaged in the design of systems in which people will work, see Taylor's ideas as part of a broad consistent view, leading from the machines to the people and treating both in one uniform way.

There are nevertheless strong (though declining) influences in the European tradition which resist a causal view when it is applied to the relations between people. We lack on the other hand any corresponding tradition to check the causal view when it is applied to the relation between ourselves and non-human nature: such a tradition for example as that expressed by the Wintu Indian woman:

> The White people never cared for land or deer or bear. When we Indians kill meat, we eat it all up. When we dig roots, we make little holes. . . . We shake down acorns and pinenuts. We don't chop down the trees. We only use dead wood. But the White people plow up the ground, pull up the trees, kill everything. The tree says 'Don't. I am sore. Don't hurt me'. But they chop it down and cut it up. The spirit of the land hates them. They blast out trees and stir it up to its depths. They saw up the trees. That hurts them. The Indians never hurt anything but the White people destroy all. They blast rocks and scatter them on the ground. The rock says, 'Don't! You are hurting me!' But the White people pay no attention. When the Indians use rocks, they take little round ones for their cooking. . . . How can the spirit of the earth like the White man? . . . Everywhere the White man has touched it, it is sore.[11]

This is an animistic view, and animism as an explanation of nature was defeated in Europe in the seventeenth century by the emergence of modern science,[12] leaving clear the way for a manipulative, causal view of non-human nature.

There is no argument in what follows for animism, which goes as far beyond the evidence as theocentric explanations of causality or purpose. What will be suggested is that, guided by the predispositions of our culture, we have constructed a causal view of the world which makes certain attitudes and actions seem natural and defensible. We can equally explain the world in terms of purpose, and quite different attitudes and actions would then seem natural.

At the end we shall have shown that the idea of a machine as a purposeless, causal device can be replaced by one which sees machines as embodying a purpose. We shall show that a programme of describing the world in terms of machines with a purpose is to all appearances just as feasible (and probably more natural) than the existing scientific programme, which attempts to show that the world is causal and purposeless. We shall suggest that this introduction of purpose allows us to bring to bear stronger arguments against some unattractive features of our technology than any which can be drawn from the current scientific and technological outlook. With its aid, we can hope to develop a different kind of technology, which co-operates with nature rather than coercing it, and which respects and responds to human purposes, rather than rejecting them.

5. *Development of the argument*

This chapter has sketched an outline of what will be attempted below. The ideas presented are often new and therefore need detailed support, which in many places would have to be mathematical. But mathematics has been entirely excluded in the main text, and the best explanation has been given which could be found without its use. Technical support will be found in the Appendices.

The accounts of control theory and of variational principles are basic to the argument, but it is by no means necessary to read them in the logical order which has been followed below. They can be omitted at a first reading, and studied when their relevance has become clear from the remaining chapters.

References

1. Joseph Needham (1927). *Man a machine*, p. 93, Kegan Paul.
2. Bertrand Russell (1979). *History of Western philosophy*, p. 181, George Allen and Unwin, reprinted Book Club Associates.
3. Eugenio Rignano (1926). *Man not a machine*, p. 10, Kegan Paul.
4. Reference 1, p. 43.
5. Reference 1, p. 45.
6. David M. Burton (1985). *The history of mathematics*, pp. 447–8, Allyn and Bacon.
7. Cornelius Lanczos (1949). *The variational principles of mechanics*, University of Toronto Press.
8. Robert Bierstedt (1957). *The social order*, p. 20, McGraw-Hill.
9. Albert Einstein (1950). *Out of my later years*, p. 114, Thames and Hudson.
10. Frederick Winslow Taylor (1906). *On the art of cutting metals*, p. 55, American Society of Mechanical Engineers.
11. D. Lee (1959). *Freedom and culture*, pp. 163–4, Prentice-Hall.
12. Reference 2, p. 522.

2 Control theory

Incorporating human purpose in machines

1. *Origins*

The engineering use of control is older than the theory by at least two hundred years. In some form or another, control can be traced back much further,[1] but a convenient starting point is the steam-engine governor. This was the first industrial use of control in a modern spirit.

James Watt's steam engines[2] were the first of their kind which could be used to drive machinery in cotton or woollen mills, or in machine shops where the looms and lathes and steam engines themselves were made. Industry was freed in this way from its dependence upon water power, and could move from the country into the towns.

Hence arose the unprecedented growth of the great manufacturing towns which was one aspect of the early nineteenth century's industrial revolution. In 1780, outside London, the largest towns were Birmingham, Bristol, and Liverpool, each with about 50 000 inhabitants, and nearly 80 per cent of the population lived in the countryside. At the 1831 census, Manchester with its cotton industry had 182 000 people, Leeds based upon wool had 123 000, Birmingham based on engineering had 144 000, while the port of Liverpool through which manufactured goods were exported had 202 000.

But there was one difficulty which could have made this development impossible. When power was taken from a steam engine, its speed fell, and when the load was removed the speed rose again. In this it resembled the water-wheel, though the changes were more severe. But in addition the speed depended upon the pressure of steam, which in turn varied with the state of the fire under the boiler. This varied more rapidly and more erratically than the supply of water to a wheel, to which its effect corresponded, and the variations in speed would have made the steam engine unsuitable for some uses, especially in spinning and weaving.[3]

A worker—a young man or a boy—was therefore stationed near the valve supplying steam from the boiler to the engine. If the speed rose, he partly closed the valve, and if it fell he opened it, keeping the speed more or less constant at the required value. Everything depended upon his vigilance and attention, and it is not in human nature to maintain these for hour after hour when the task has no inherent interest. Workers in the early twentieth century, carrying out a similar task in the chemical industry to control a temperature, were provided with a one-legged stool. They could rest, but could not sleep.

James Watt did not always respond with enthusiasm to the demands made upon an inventor: he once exclaimed that of all things in the world, nothing is more stupid than inventing. But faced with the problem of controlling the speed of his engines, he produced an outstandingly successful solution. It was not entirely original, because it had been used earlier for controlling the distance between the upper and lower grindstones in a windmill.[4] But technically this was very different from the application made by Watt: 'open-loop' rather than 'closed-loop' in terms that will shortly be explained. The difference is profound, and Watt himself was probably not fully aware of its significance.

The 'fly-ball governor', or Watt's governor, is sketched in Fig. 2.1. Two large heavy weights are hung on arms, pivoted on a rotating vertical shaft driven by the engine. As the speed of the engine changes, the weights move: out and up when the speed rises, down and in when it falls. By two links, the weights raise or lower a collar on the rotating shaft, and the movement of the collar is transferred by a lever to the steam valve.

When the speed rises, the collar also rises, and the steam valve is moved towards the closed position. When the speed falls, the steam valve is opened. The governor therefore does automatically what the human attendant was required to do, in order to prevent large changes of speed.

Can this device keep the speed exactly constant? A little thought shows

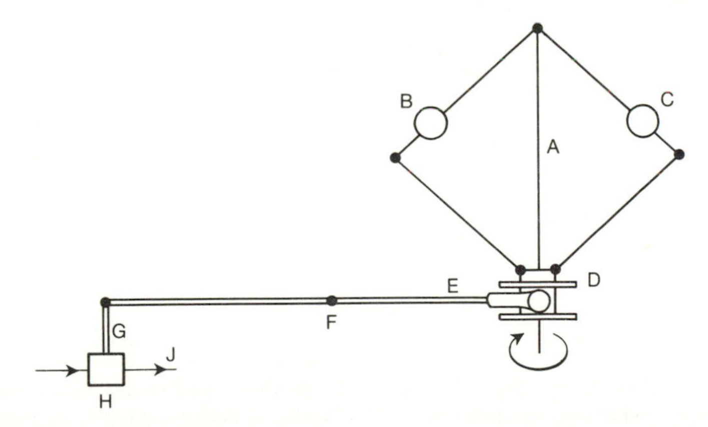

Fig. 2.1 The vertical shaft A rotates at a speed proportional to the speed of the engine. The fly-balls B, C move out and up as the speed increases, raising the collar D. This in turn lifts the end of the lever E which is pivoted at F. The rod G moves down, causing the steam valve H to restrict the flow of steam J to the engine. As F is moved to the right, the motion of the collar needed to make H go from fully open to fully closed is reduced.

that it never can do so. When the engine is loaded, it requires more steam than when it is running light. The steam valve must be further open, and the collar on the governor must be lower down. In turn, the fly-balls must be lower, and so the speed must have fallen: the amount by which it falls was once called the 'droop', but is now referred to as 'offset'.

The extent to which the collar must move down (and hence the fall in speed) can be changed by moving the pivot point of the lever which connects the collar to the steam valve. As the pivot is moved to the right, the travel of the collar needed to move the steam valve from fully open to fully closed is reduced. Hence it seems that the constancy of speed can be improved continually, within limits set by friction and other forces, by moving the pivot towards the collar.

What happens in practice is different, and surprising. If we start with the pivot far from the collar, moving it towards the collar does indeed improve the constancy of the speed. When the pivot reaches a certain point, however, something new begins to happen. At each change of load, the speed oscillates above and below its final value before settling, as in Fig. 2.2. Further movement of the pivot makes these oscillations more pronounced, and when the pivot is moved still nearer to the collar, the system becomes unstable: the oscillations no longer die away, but increase spontaneously until the physical limitations of the engine prevent their further growth. The speed is said to 'hunt', and the extent of the speed changes may be so large as to be destructive; they will at all events make the engine unusable.

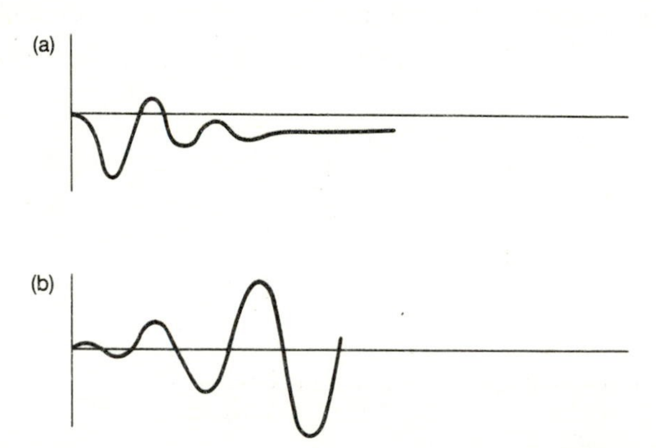

Fig. 2.2 As the pivot F in Fig. 2.1. is moved to the right, there comes a point where an increase in load causes the speed to oscillate as in (a) before settling to its new lower value. Further movement of the pivot to the right causes a spontaneous build-up of oscillation as in (b): the system is unstable, and the oscillations in speed ('hunting') increase to an extent which is only restricted by physical limitations of the engine and governor.

2. *Theory*

It is at this point that theory must begin, though in fact the first analysis in a recognizably modern spirit was not made until 1868, and it required the attention of the greatest physicist of the day, James Clerk Maxwell. Meanwhile, engineers developed an empirical understanding and successfully applied governors to their engines, while inventors pursued a largely unrewarding search for the 'isochronous governor', which would keep the speed constant regardless of the load. The inventors of such a device, as Maxwell drily remarked, 'naturally confine their attention to the way in which it is *designed* to act,'[5] whereas the difficulties arose from a kind of behaviour that was unexpected and unexplained.

Let us look at the matter from a modern point of view. We now distinguish sharply, in a way that Watt or Maxwell did not, between things in which we are interested only as information, and things which serve some other physical purpose. The speed of the engine is information, and we can represent it indifferently in a multitude of ways. It can be represented by the position of the heavy fly-balls of a governor, or by a voltage generated by a small electrical generator fixed to the engine. Or we can fix a disc to the shaft of the engine, in which there are a number of holes, and by shining a light through the holes can generate an electrical pulse at, say, every 30° of rotation. Then by digital circuitry we can measure the time between successive pulses and calculate the speed some thousand or more times a second. All of these representations require some power, but this is irrelevant to the information and has been continually reduced by the progress of technology.

On the other hand, the physical flow of steam to the engine, and the rotation of its output shaft, have a significance beyond any measurement we can make. The steam carries energy, and has to be generated by a large boiler consuming fuel. The rotation of the output shaft drives machinery and transfers to them some of the energy contained in the steam.

Information about physical variables such as speed is obtained by a measuring device which we call a 'sensor'. To make use of the information we use an 'actuator' which accepts information and acts in a corresponding way on the system. Then we can represent the steam engine and its governor in a schematic way by Fig. 2.3.

There is a steam flow to the engine and an externally imposed load. As a result of these the engine runs at a certain speed—we think of the steam flow as an input, the load as a disturbance, and the speed as an output. The governor develops a 'signal' (the fly-ball position) carrying information about the speed. This information is transmitted to the actuator, which is the steam valve, and changes the steam flow.

Looked at in this way, it is clear that the engine is given, but the

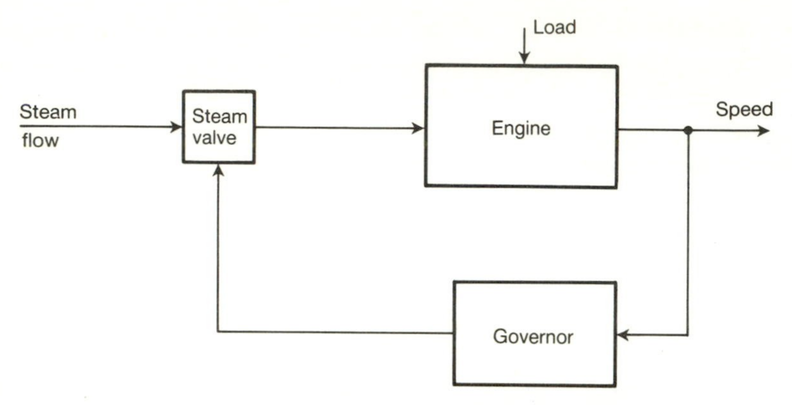

Fig. 2.3 We represent the engine as a 'black box', which accepts a flow of steam from the valve and rotates at a speed determined by the steam flow and the load. The governor has the speed of the engine as its input, and produces a 'signal' (here the movement of a lever) to the steam valve. The existence of a closed loop is obvious. In principle, many other ways could be found of generating a signal from the speed and causing this to change the steam flow.

remaining parts of the system could be changed in a multitude of ways without fundamentally altering what goes on. The sensor could generate an electrical voltage, this could be amplified, and applied to an electric positioning motor which operated the steam valve. Alternatively, we could generate electrical pulses to represent the speed, and process these in a digital computer, before generating an electrical signal as before to operate the steam valve by means of an electric motor.

The essential feature in Fig. 2.3, and the deep significance of Watt's invention, lies in the fact that there is a complete loop. Steam flowing to the engine with a given load produces a given speed. This results in a certain fly-ball position, and a certain steam valve position. Consequently there is a certain steam flow, which results in a certain speed . . . and so on. There is no beginning and no end: we have to consider everything at once. And we have to do this not only when the engine is operating steadily, but also when changes are brought about, for example by alterations in the load or the steam pressure.

We say that the complete system in Fig. 2.3 is a 'closed-loop control system', or a 'feedback system'. We can break the loop, for example by removing the lever which connects the collar on the governor to the steam valve. Then a change in the steam valve setting, with a given load, will alter the speed, which will change the fly-ball position—but no further influence will be transmitted to the valve. The system is 'open-loop', and its behaviour will be quite different from before, and much simpler to analyse.

The difference between an open-loop and a closed-loop system may have been apparent to Watt intuitively, but there was no general theory to provide an understanding of its consequences. When the fly-ball device was used to change the distance between the upper and lower millstones of a windmill for grinding corn, as the speed changed in response to the wind, the system was open-loop. Changes of speed changed the gap between the stones, but this did not appreciably affect the speed. If the fly-balls had been used to feather the blades in order to keep the speed constant,[6] a closed-loop system would have been created. In applying the fly-balls to control the speed of his steam engine, Watt created a closed-loop system and began the development which has led to modern control theory.

3. *Designing a control system*

Maxwell provided a foundation for the analysis of control systems, but it was a long time before much further progress was made. The reason lies in the difference between analysis and design. Given a control system, such as Watt's, which has already been designed, we can analyse its behaviour. We use the principles of physics (and of chemistry or whatever else is needed) to write down the mathematical equations which govern its behaviour. Most of this will already have been done for us by engineers concerned with the separate parts of the system: steam valves, steam engines, chemical reactors, or whatever else. Then putting all the equations together we can solve them, if necessary by using a digital computer. In this way we can find how the system will behave.

The case is different if the system does not yet exist, and we want to design it. We can, of course, generate a design without any theoretical guidance, and then analyse it to see if it is satisfactory. But if it is not, what do we do to improve it? Watt, with his intuitive understanding, made his governor work without any analysis. Subsequent inventors aiming at an isochronous governor which would keep the speed exactly constant were generally less successful. Maxwell showed some of the reasons for their difficulties, but gave no general way of producing systems in which the difficulties could be overcome.

The development of effective design methods began in the 1930s, but a number of empirical developments[7] took place during the early 20th century. Together with the mathematical analysis which accompanied them, these laid the foundations on which this progress was made. Some of the most significant milestones are the following.

i. In the second decade of the century, ships' steering engines were analysed by Minorsky. In smaller vessels, the wheel was connected directly to the rudder. In large steamships the forces were too great to allow this: 'in some of the armour-plated ships of the British Navy it requires nearly a hundred men to put the helm hard over when the vessel is going at full speed.'[8] From the mid-nineteenth century, therefore, a steam engine was used to move the rudder, in response to demands from the wheel. The wheel is turned, giving a desired position for the rudder. The steering engine moves the rudder to eliminate the discrepancy between its actual position and the demanded position, in the same way as power steering in a car. The major function of the system, shown diagrammatically in Fig. 2.4, is to respond to demands from the wheel, and it is therefore called a 'servo system'. The Watt's governor, in which the desired speed is rarely changed and the aim is to counteract load changes, is called a 'regulator'.

ii. The early chemical and petroleum industry, from the 1920s onward, created a strong demand for control systems ('regulators') to maintain desired conditions of temperature, pressure, flow, liquid level, and other process conditions. Instruments were developed to measure these variables, and the information was transmitted as an air pressure in the range 3–15 psi, illustrating an earlier comment that information can be conveyed in quite arbitrary forms. Here electrical transmission of the information would have been possible but in many situations the risk of explosion from sparks ruled it out with the technology of the day.

The information about a controlled variable was conveyed as an air pressure through small copper pipes to a controller, as Fig. 2.5. This compared the measured value with the desired value, and sent a correcting signal to a pneumatically operated valve, which changed a flow in order to obtain the correct condition: the flow in temperature control, for example, might be a flow of steam for heating.

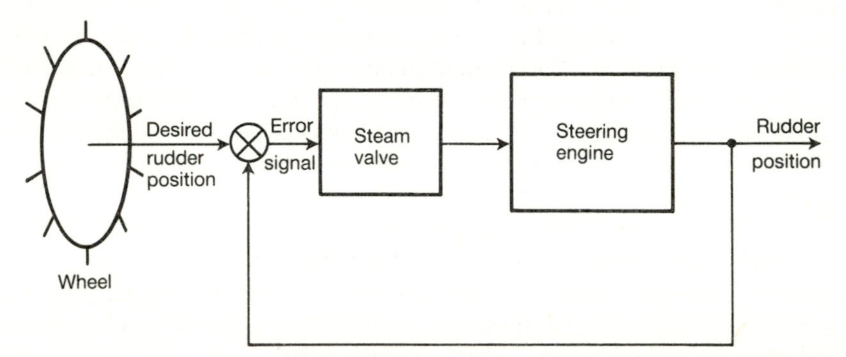

Fig. 2.4 The ship's wheel transmits a desired rudder position which is compared with the actual position. The difference between the two (the error signal) causes the steam valve to move in a way which will reduce the error. The rudder is therefore driven (very nearly) to the position which has been demanded from the wheel. Notice that the steam valve is more complicated than in Fig. 2.3, because it can now cause the engine to rotate in either direction.

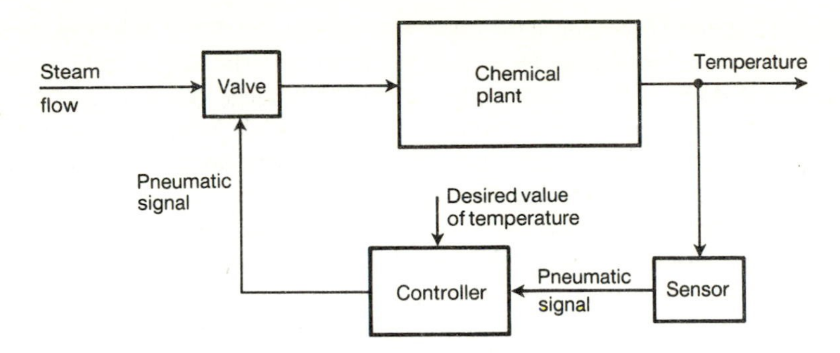

Fig. 2.5 Pneumatic control system for a temperature in a chemical plant. The sensor produces an air pressure in the range 3–15 psi as the temperature changes through the range for which the sensor is designed. This air pressure is transmitted through a small copper pipe to a controller, which is often in a control room some distance from the sensor. In the controller an error signal is obtained by comparing the measured and desired temperatures, and the error signal may be modified to improve stability and reduce offset. The controller often displays and records the temperature. A signal from the controller, again in the range 3–15 psi, operates a valve to change the flow of steam for heating.

A chemical plant would have a multitude of control loops for pressures, temperatures, flows, levels, compositions, etc. In a modern plant the information is transmitted electrically, and stringent means are used to limit the energy so that sparks will not ignite any inflammable gas which may be present. All the controllers are now often replaced by a single computer which samples the variables in turn, very rapidly, and produces the appropriate signals to the valves.

An empirical art of using these instruments and controllers became highly developed, and there were standardized methods for improving stability ('derivative action'); and also for eliminating offset ('integral action'), so that they were the equivalent of an isochronous governor. But of a general design theory for these systems there was as yet no sign.

iii. From the early 1920s onward, the telephone system spanning the United States was being rapidly developed. The speech information was conveyed as an electrical voltage along copper wires, which initially were very heavy and expensive. There was a strong economic incentive to reduce their size, but when this was done the electrical signal was rapidly reduced in magnitude as it travelled down the line.

After some tens of miles, depending on the size of the conductors, it became necessary to use an electronic amplifier, based at that time upon valves, to increase the signal before it was fed to the next section of line. Any distortion or variability in an amplifier was compounded by the large number through which the signal passed. Telephone engineers therefore faced the problem of producing an amplifier with much more stable properties, and much less distortion, than anything previously available.

Black,[9] at the Bell Telephone Laboratories, solved this problem empirically by inventing the principle which is still used in the hi-fi amplifier, Fig. 2.6. The output of the amplifier is divided by a large number, say 100, which can be done with extreme accuracy. The resulting voltage is subtracted from the incoming signal and any discrepancy is passed to an amplifier. This multiplies the discrepancy (the 'error') by a very large amount, say 100 000, though with some distortion. The output of the amplifier drives the outgoing voltage in the direction which will reduce the 'error', which is the signal at its input.

Now the output of the amplifier is about 100 000 times the input, so the latter must be very small. But the input, the error, is the difference between the incoming signal and exactly one hundredth of the outgoing signal. Hence the outgoing signal is, with very good accuracy, 100 times the incoming signal. The accuracy depends on the elements in the feedback path which divide by 100, and these, which are just resistors, are accurate and stable. So long as the amplification in the forward path is high, its exact value, and the distortions which accompany it, have little effect.

Black's amplifier worked wonderfully well, but the same problems which dogged the Watt's governor arose again to trouble it. The greater the amplification used, the better we expect the performance to be. One million should be better than 100 000. But it is a closed-loop system, and as with the governor there comes a point, as we increase the amplification, where the system becomes unstable. Black himself appreciated this problem, and developed an early version of the design methods which would be necessary to deal with it. But no complete understanding was available at the time.

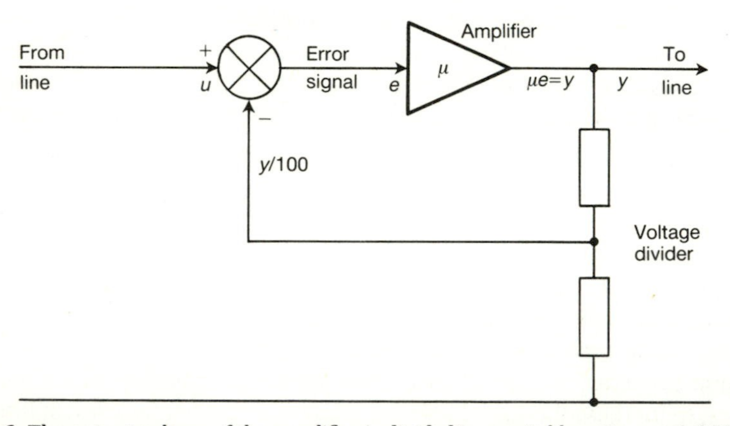

Fig. 2.6 The output voltage of the amplifier is divided in a suitable ratio, say 1:100, and compared (simply by subtraction) with the incoming signal. The difference (the error signal) is amplified, driving the output of the amplifier in the direction which reduces the error. Thus one-hundredth of the output voltage is made very nearly equal to the input voltage. An elementary analysis is given in Appendix 1.

Following the work of Maxwell and those (particularly Routh) who extended it, there was no great problem in saying whether an amplifier, already designed, would be stable or unstable. But the engineers at Bell Laboratories went further. Following the lead given by Black, graphical methods of analysing a system were developed by Nyquist and Bode, and others later, which showed whether it would be stable. Their methods did not stop at this point—if they showed that the system was unstable, they suggested how it could be stabilized. If it was stable, but insufficiently so, they suggested how stability could be improved. They showed how offset could be reduced, and what effect this would have upon stability. For the first time, it became possible to design closed-loop control systems in a consistent and logical way.

At this point, the story goes underground. During World War II, the methods of Nyquist and Bode and their colleagues were refined and extended and applied at MIT, under the pressure of military demands. Servo systems were designed which would point a radar antenna accurately in a specified direction. By a further development, once the antenna had been pointed towards an aircraft, it would continually follow it, measuring its direction and distance. Then this information was fed to a 'predictor' which was an electromechanical device: this was before the days of the digital computer. The predictor specified where anti-aircraft guns should be pointed, how the shells should be fused, and when they should be fired, so that the shells would explode at the predicted position of the aircraft. This information was fed to servos controlling the pointing of the guns, and the whole process was thus automated.

After the war, the accumulated experience at MIT was published,[10] and the design methods, now highly refined, were applied to a wide range of industrial problems. The methods relied upon pencil and paper and slide-rule, and what could be done with these means was more or less accomplished by 1960. Later, from about 1970, it became possible to use a digital computer with graphical output to replace the pencil and paper and slide-rule, and the methods were further extended, and applied to new problems.[11] But in the 1960s there arose quite different methods, devised for a different kind of problem.

These new methods, described at the time as 'state-space methods', or 'modern control theory', marked a decisive divergence of view which will be discussed in more detail later. The divergence occurred also to some extent in other branches of engineering, and was associated with the growth of 'engineering science', but its influence in control theory was probably greater than elsewhere. Before discussing its implications, let us look at its technical content.

4. *Synthesis*

Analysis, as we have described it, starts from a system which already exists, or which at least has been already designed, and deduces mathematically the way it will behave. The theorist can then propose to invert this process. Let us specify what we want our system to do, in such detail that only one system will satisfy our requirements. Then it becomes, in principle, a mathematical problem to find this system. The process of doing so is called synthesis, and it is tempting to equate synthesis with design, or to say that synthesis is what design ought to be if it were fully understood and systematized.

The connection between synthesis and design will be taken up later (in Section 7), and we shall first follow the course of the idea in control theory. It began with work by the mathematician Norbert Wiener[12] during the war. He considered the problem of separating a signal from noise. For example, a radar set sends out a pulse of radio waves, and their echo returns. But the echo is faint, and all kinds of random disturbances (which we call noise) are superimposed on it. The distance of the object producing the echo is determined by the time between transmission and the receipt of the echo, but the echo is ill-defined because of the added noise. We wish to separate, as well as we can, the signal (that is, the echo) from the noise.

If signal and noise had exactly the same properties, we should be unable to separate them, but usually there are some differences: often the noise predominates at high frequencies and the signal at low frequencies. The situation is the same as when we listen to a distant radio transmission, or an old gramophone recording. Superimposed on the music is a random disturbance, much of it sounding as a hiss. If we reduce the reproduction of high frequencies, by means of the tone control or a special filter, we lose some components of the music, but we cut out proportionately much more of the noise. Our filter makes an approximate separation of signal and noise, accepting the former and rejecting the latter.

Wiener posed a problem of this kind in mathematical form: given the properties of the signal and of the noise, and a criterion for judging the goodness of the separation, what is the absolute best filter to use? By some delicate mathematical procedures, which he had himself helped to develop in earlier years, he was able to give a solution.

The main developments in control, however, were of a different kind, and were associated with Bellman, Kalman, Pontryagin, and many others.[13] A typical problem was this: given the aerodynamic characteristics of an aircraft, the behaviour of its engine, and the properties of the atmosphere, how should it be manoeuvred in order to climb to its operating height in the minimum time? The answer,[14] for a particular aircraft, turned out to be different from what might be expected, Fig. 2.7. The best procedure was to

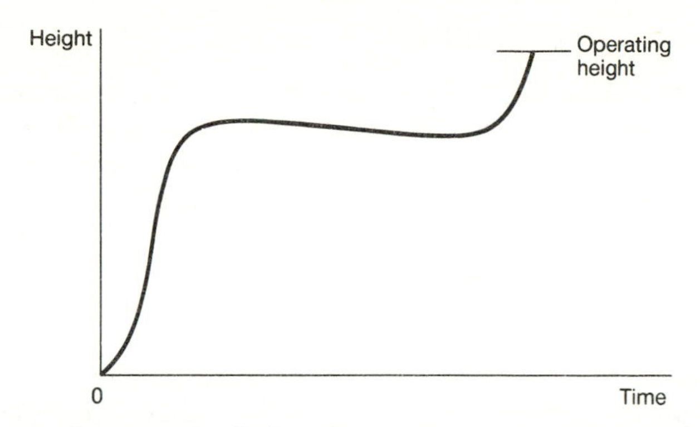

Fig. 2.7 An aircraft is required to climb in the minimum time to its operating height. This is achieved by a steep initial climb, followed by a shallow dive to gain speed, and a final climb to height. Any deviation at any point from this schedule will increase the time.
The calculation is for a particular aircraft under particular conditions, but the result is shown only in schematic form.

climb steeply to some particular intermediate height, then go into shallow dive to build up speed, and finally pull up again into a climb. The problem, in fact, lies within the scope of the 'calculus of variations', which originated with the Bernoullis before 1700, but is here approached from an engineering perspective.

Another problem in a similar vein would be the following. We have an industrial plant such as an oil refinery in which the best values of the temperatures, pressures, flows, etc., are known. There is a cost at each moment for departing from these values, and there is a cost for applying control to bring the variables back to their best values. How should we operate the plant in order to incur the least cost over some future period? Or again, we have a rocket on the launching pad, and wish it to reach a certain point in orbit at a certain time, with a certain velocity: and to do all this with the minimum consumption of fuel. How should the rocket be manoeuvred to achieve this? In all these problems it is usual to refer to the quantity to be minimized as the 'cost', though it need not be expressed in monetary terms, and in one of our examples, illustrated by Fig. 2.7, it was a time.

All of these are problems of synthesis, and their study led to important advances in the theory of control. Typical problems were those of 'existence and uniqueness'. That is, under what conditions on the system can I know that a particular specification for its behaviour will actually lead to a solution? And under what conditions, when there is a solution, will there be only one? The most important development for our later discussion, however, is a technique introduced by Richard Bellman.

5. *Dynamic programming*

Bellman's technique[15] of 'dynamic programming' is adapted to the kind of synthesis problem we have just discussed. It can be used to obtain numerical solutions to some problems, but its main interest for us will be conceptual. We shall illustrate it by the problem of navigation which we considered in Chapter 1, omitting here any reference to random disturbances.

We have, therefore, a boat which we wish to sail from a given point A to another given point B, Fig. 2.8, in a given time T. We know exactly the direction and strength of the tide at every point during this time. We also know how the fuel consumption, and the speed through the water, vary with the speed of the engine. Because there is a fixed relation between speed of the engine and speed through the water, we can regard the speed and heading of the boat as the variables we have available for control: the speed and the heading (that is, the direction of travel with respect to the water) are together called the 'velocity'. We now ask, how should the velocity be chosen at each instant, so that we go from A to B in the time T, and do so while using the least possible amount of fuel.

For definiteness, let us say that the fuel consumption at each moment depends on the size of the boat—that in fact it is proportional to the displacement, so that doubling the displacement doubles the required consumption at a given speed. Let us also say that the fuel consumption depends on the square of the speed—doubling the speed quadruples the consumption. Finally, to simplify matters at the beginning, let us say that there is no tide, so that the speed through the water is also the speed with respect to the land.

We notice first that we must have some condition besides the fact that we

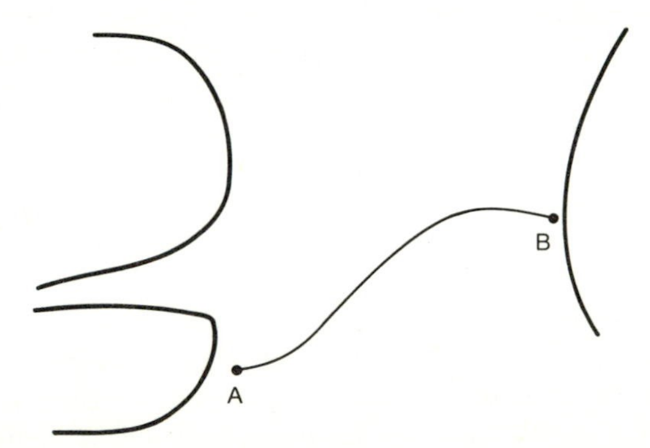

Fig. 2.8 A boat is to be sailed from a given point A to another given point B. The starting time and finishing time are given, together with the flow of the tide. How must the boat be navigated in order to accomplish the journey in the given time, while using the least amount of fuel?

start from A, otherwise a solution would be to remain at A and use no fuel at all. We have said here that we go to a fixed point B, but other conditions are possible (such as the initial position and initial velocity) and will be used later. We have also specified the time available, because otherwise, in still water, the slower we went the less fuel we would use: halving the speed would double the time, but reduce the consumption at each moment to a quarter, so the total fuel used would be halved. Hence as the time increased, the fuel used would continually decrease, and there would be no speed which gave the minimum consumption of fuel. (The speed zero does not qualify because it will not take us to B, while if we propose any speed greater than zero this cannot be the best: half the proposed speed will need less fuel.)

The property which characterizes the solution to this problem, when it is defined in a satisfactory way, is easy to state. If we knew the solution, that is the best speed and heading at each instant, then any change which we made, at any time during the journey, either from the best speed or the best heading, would increase the fuel consumption no matter what we did at other times. This is enough, mathematically, to solve the problem, but it gives us singularly little insight into the form of the solution.

What Bellman pointed out is that the problem can be solved very easily provided that we have one thing. What we need to know is the minimum amount of fuel needed to get to B in the remaining time: we need to know this for each time during the journey, and for every point we might reach (whether or not it is on the best path). 'Best' is usually replaced here by 'optimal', and the information just described is the 'optimal cost function'. We shall call it for short the OCF.

Knowing the OCF we argue in the following way. We know where we are the present time, say at C, Fig. 2.9. From here, in some short time interval such as a minute, we can reach a wide range of different points by sailing in different directions at different (constant) speeds. To reach one of these points, say D, will cost us a certain consumption of fuel, which we can easily calculate. Then if we add the OCF at D for the time one minute from now, this is the fuel we shall need to sail to D in the next minute, and then to sail in the optimal way during the rest of the journey to B. By comparing the figures obtained in this way for all possible points D, E, F, G, etc., we can easily (in principle) pick out the best point to reach in the next minute. This determines approximately the best course and speed, and if we successively reduce the interval from one minute we shall get better and better approximations.

At first sight this seems like a circular argument, since to get the OCF at D it appears that we need to know the best course starting from D at a specified time. So we can solve the problem if we already know the answer. Actually, things are not as bad as this. If we write down mathematically what we described in the previous paragraph, we can find an equation to solve for the

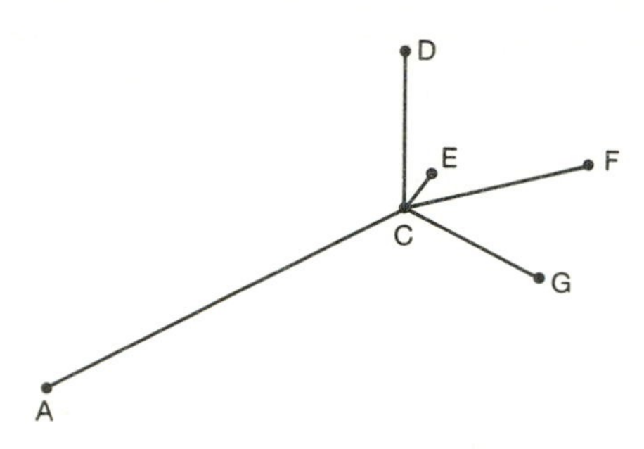

Fig. 2.9 Our boat is at C, and by sailing in different directions at different (constant) speeds for one minute we can reach a wide range of different points such as D, E, F, G. It is easy to calculate how much fuel is needed for this. We also know the least amount of fuel needed for this. We also know the least amount of fuel needed to get from each of these points to B in the remaining time (this is the OCF). We can therefore compare the relative advantages of going to D, E, F, G, etc., and pick the best. Then we can repeat the procedure from this new point, and so on.

OCF—it is known as Bellman's equation. In general, the form of the equation is complicated, but then we are solving a complicated problem, so this is not surprising.

Once we have the OCF, finding the best path can in fact be quite easy in particular problems. For our boat, we assumed that the fuel consumption depended on the square of the speed. It then follows that the best course is always perpendicular to the line of constant OCF passing through our present position at the present time. The best speed is proportional to the 'gradient' of the OCF—that is, its rate of change with distance along the course.

If we think of the contours of the OCF at a given time as defining a hill, both these statements are summed up by saying that we must go downhill in the steepest direction (perpendicular to the local contour) at a rate proportional to the slope. For our excessively simple example, the appropriate action is self-evident: we sail at the constant speed in the constant direction which takes us from A to B in the given time. The way this solution arises from the OCF is illustrated in Fig. 2.10.

Sailing the boat from A to B in the given time could be accomplished by human action—by the captain in person steering the boat and adjusting the engine speed. Let us suppose, however, that we have devised a control system to accomplish the aim automatically. There are then two ways in which we can arrange this control system.

i. Before we start, we can work out the OCF. As explained above, this defines the optimal speed and heading when we know the time and our

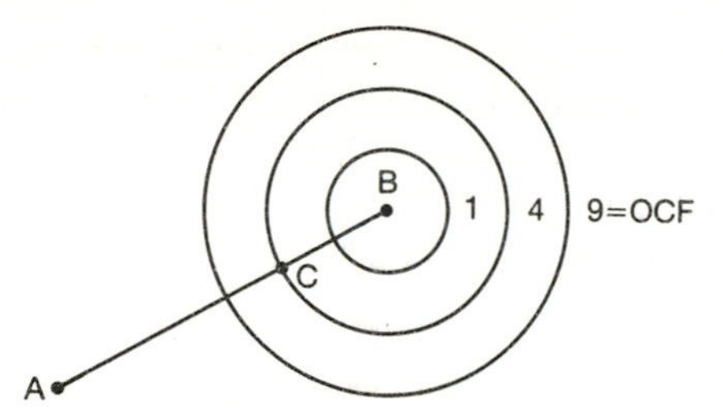

Fig. 2.10 The contours of constant OCF are circles centred on B. The values of the OCF at a particular time, when our boat is at C, are shown: they increase as the square of the distance from B. This is because the remaining time is the same wherever we are. Therefore the necessary speed is directly proportional to our distance from B, and the fuel consumption goes up with the square of the speed. We think of the contours of OCF as defining a hill (or as here a dish-shaped hollow) and go downhill in the direction of steepest slope, which leads directly towards B.

At other times, the contours of the OCF will still be circles around B, but they will have different values associated with them.

position. But as there are no random disturbances, we can obtain our present position at any time by dead reckoning, knowing our speed and heading over the time since we started from A.

We therefore measure our speed and heading continuously, and feed them to the controller. It calculates the best present speed and heading, and adjusts the engine and the rudder accordingly. We have a closed-loop system (implementing a policy as defined in Chapter 1) according to the scheme in Fig. 2.11.

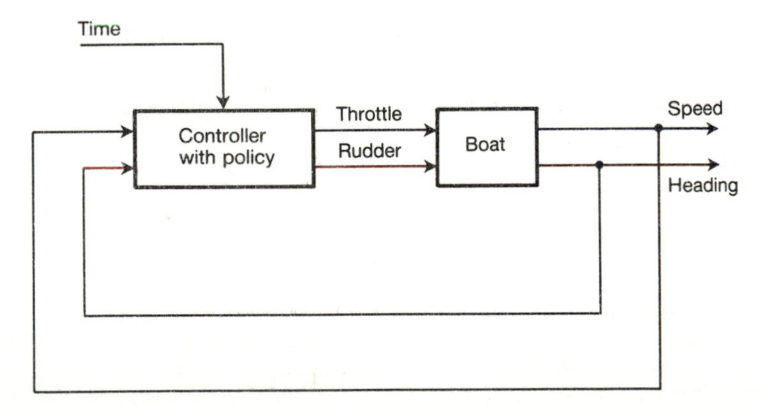

Fig. 2.11 Control system to implement a policy. The controller contains the policy, which gives a relation between position and time, as causes, and speed and heading as effects. Position can be obtained from the speed and heading during the time since we started (there are no disturbances and we are in still water). The closed-loop controller manipulates the throttle and steering to produce the optimal speed and heading during the voyage.

ii. Because there are no random disturbances, what will happen is entirely predictable before we start from A—we can work out beforehand what our best speed and heading will be at each time. Then our controller can be supplied with this information, and can adjust the engine and rudder accordingly, knowing only the time, Fig. 2.12. This implements a schedule.

The difference between the policy and the schedule here is only in the way we regard the problem and implement the solution: both produce the same result. But it will already be clear that if there are random disturbances, and if we are able to find our position by some independent means, the policy is likely to be better than the schedule. We shall return to this point in Chapter 5.

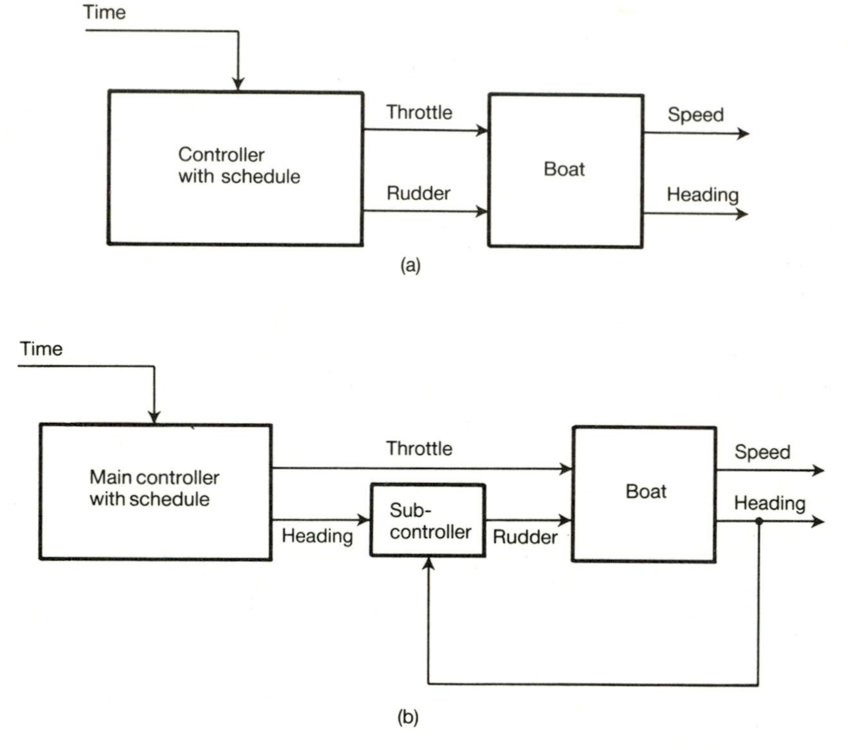

Fig. 2.12 System to implement a schedule. The open-loop controller is supplied only with the time, and contains a pre-computed schedule which gives the speed and heading at each time, and the throttle and rudder settings needed to produce them. The throttle and rudder are manipulated by the controller to produce the required speed and heading, as in (a).

Practically, this would not be workable, because the relation between heading, and movement of the rudder, cannot be predicted with sufficient accuracy. A sub-loop would therefore be used as in (b), but the main controller remains open-loop. A similar sub-loop could also be used to control the speed.

6. *Purpose and causality*

We have mentioned three possibilities above: the captain can sail the boat himself, or we can design a control system to apply a policy, or we can design the controller to implement a schedule. These three possibilities have three different interpretations in our ordinary way of understanding purpose and causality.

i. We should say, in ordinary speech, that the captain has the purpose of going from A to B in the given time, while using the least amount of fuel. He carries out this purpose by adjusting the engine and rudder in the appropriate way.

ii. We should say that the engineer who designed a controller to implement a schedule had the aim of causing the boat by itself to accomplish the purpose stated in (i): that is, his aim was to incorporate the human purpose in a machine. He did this by deciding beforehand what would be necessary at each time during the voyage, and building this into the controller.

iii. The engineer who implemented a policy would again be said to incorporate the human purpose in a machine. But this time he does so in a different way. He does not predict where the boat will go, but gives a general rule for it to obey. The rule says, in effect, if the position of the boat is such, and the time is such, then the optimal speed and course are such and such.

The last statement defines a relation of cause and effect, so that a policy translates a purpose into the causal relationships which will accomplish it. The effect, which is here the speed of the boat and its course, is made to depend only on present causes such as the time, and the position of the boat, and this requirement is built into the mathematical procedure by which we make the translation.

In the next chapter we shall explain how Hamilton succeeded in deriving Newton's laws of motion from a purpose. What he showed was that the policy which implemented the purpose was identical with the causal relations defined by Newton. Hamilton's principle, which incorporates this result, has been profoundly important in the subsequent development of physics, but its philosophical implications have always been suspect in orthodox science.[16]

The objection raised is that if a system—say a stone moving under the influence of gravity—is said to obey a purpose, this implies that it foresees the future. It is envisaged that the purpose will be achieved by means of a schedule. Then at the initial time, the stone must know the future path it is to follow, depending upon its initial position and velocity. At later times it must remember this path and follow it.

The objection is not that the schedule anticipates the future in the sense of making actions at one time depend upon information which can only be available at a later time: the 'cause' is the combination of initial position and velocity, and this determines the path at later times. The difficulty lies in the delay between cause and effect, so that the latter becomes known before it occurs, and has to be remembered until it is needed. As the language of 'knowing' and 'remembering' indicates, these ideas are entirely foreign to the usual understanding of inanimate objects.

The difficulty is to a large extent artificial, arising from our assumption that there are no random disturbances whatever. The initial conditions, as Laplace remarked,[17] then determine the whole course of future events. When we add random disturbances, as we shall later do, the matter becomes much clearer and more natural.

But even in the absence of disturbances, there is a complete answer to the objection. We can achieve the purpose by means of a policy just as well as by a schedule; and in the absence of disturbances the two are equivalent, producing exactly the same effect. The policy is a relation between a cause and an effect at the same instant, conforming to the usual description of physical laws such as Newton's equations. It gives the relations of cause and effect which accomplish the purpose incorporated in Hamilton's principle.

If this reference to purpose were rejected, as it is in orthodox science, we should not be able to say that any machine embodied a purpose. In (iii) above, we spoke of a human purpose being incorporated in a machine. The purpose was first translated into the causal relations which would fulfil it, and the machine was built to obey the causal relations. The purpose has now disappeared, and the machine is a causal device. Did the purpose ever exist?

Common sense says that it did, but there is a strong tendency in science to say it did not. The human body is a physical system, and in principle it can be explained as a causal device. Then, as with a machine, purpose disappears: 'In science, man is a machine; or if he is not, then he is nothing at all.'[18] We shall return continually, and from different directions, to this theme.

7. *The control engineer*

The account which has been given of the history and practice of control is merely a sketch, sufficient for the present purpose: more complete accounts of the earlier history are given elsewhere.[1,3] The subject has a deep interest in itself, and underlies nearly all of our present technology. At one extreme, control theory is a branch of applied mathematics giving scope for the application of a surprisingly wide range of techniques: functional analysis, calculus of variations, algebraic geometry, stochastic analysis, and many others. At a different extreme it deals with practical systems and their engineering design, usually with the aid of computers and often with graphical output and interactive working.

An associated discipline is decision theory, which is concerned with similar problems, but aims to guide human actors, rather than to design machines which act on their behalf: it is therefore less concerned with the technicalities of implementation. An extension of the ideas of control leads to games theory: in control only one purpose is envisaged, our own, and this is to be imposed upon the natural world which is seen as inert and purposeless. In games theory, two or more competing purposes are considered, and we wish, for example, to do the best we can to achieve our purpose while our adversary is doing the best he can to frustrate it.

All of these have in common a concern with some underlying human purpose: if there were no purpose there could be no control. At the same time, control engineering is embedded in the scientific and technical endeavour from which all references to purpose are excluded. The underlying conflict surfaces from time to time, and when it does the practising control engineer is perhaps more willing to accept the reality of purpose than those with a different experience. At the same time he has, as will be shown later (in Chapter 5, Section 10), more reason to distinguish between human purpose and the behaviour of machines into which this purpose may have been incorporated.

One important point in control engineering where the conflict between purpose and causality surfaced was in the relation between synthesis and design. The theoretical advances which arose from such methods as optimal control, based upon synthesis, were easy to accept. So too was the guidance which could be obtained in some special problems such as the minimum-time-for-climb of an aircraft. Much more debatable was the implicit claim that synthesis could replace design.[19]

Synthesis is carried out in a computer, and in principle there is no human intervention. The designer simply says what he wants, and the computer deduces the system which will satisfy his requirements. If we object that designers do not work in this way, but develop the requirements and the design to some extent concurrently, as they discover what is actually possible, then we situate ourselves at once in a continuing debate. It is the debate at one level about engineering as applied science, versus engineering as a practical art.[20]

At another level it is the debate in 'artificial intelligence' about whether there is anything which a human being can do, which cannot also be done by a computer.[21] Stated in a rather crude form, the argument which denies that any such thing exists runs as follows. A man is a machine. The computer is a universal machine which can simulate any machine. Therefore anything which can be done by a man can also be done by a computer, given suitable sensors and actuators to connect it to the surrounding world. All that is necessary is to specify precisely what a man does when carrying out a task (such as designing a control system) and a computer can be programmed to do it.

To argue against this position within the scientific framework is very difficult, because it amounts in appearance to arguing that a man is not a machine, and is therefore (at least in some aspects) outside the scope of scientific study. What will be argued here is something different: that man is a machine, but not the kind of machine without purpose which is usually implied in the statement 'man is a machine'. The sense in which this statement is to be taken will become clearer as we proceed.

References

1. Otto Mayr (1970). *The origins of feedback control.* MIT Press.
2. Robert E. Schofield (1963). *The Lunar Society of Birmingham*, pp. 333–5, Oxford University Press.
3. S. Bennett (1979). *A history of control engineering 1800–1930*, p. 13, Peter Peregrinus.
4. Reference 3, pp. 10–13.
5. J. C. Maxwell (1868). On governers, *Proc. Roy. Soc.*, vol. 16, pp. 270–83; reprinted in Richard Bellman and Robert Kalaba (editors) (1964). *Selected papers on mathematical trends in control theory*, pp. 3–17, Dover.
6. Reference 3, pp. 10–13.
7. Reference 3, pp. 142–7.
8. J. MacFarlane Gray (1867). Description of the steam steering engine in the *Great Eastern* steamship, *Proc. I. Mech. E.*, p. 267; quoted in Reference 3, p. 98.
9. Reference 3, pp. 190–5.
10. H. M. James, N. B. Nichols, and R. S. Phillips (1947). *Theory of servomechanisms*, McGraw-Hill.
11. H. H. Rosenbrock (1974). *Computer-aided control system design*, Academic Press.
12. Norbert Wiener (1949). *Extrapolation, interpolation, and smoothing of stationary time series*. John Wiley.
13. R. E. Bellman, I. Glicksberg, and O. A. Gross (1958). *Some aspects of the mathematical theory of control processes*, Rand Corporation.
14. A. E. Bryson and W. F. Denham (1962). A steepest-ascent method for solving optimum programming problems, *Trans. ASME, J. Applied Mechanics*, June, vol. 84, pp. 247–57.
15. R. E. Bellman and S. E. Dreyfus (1962). *Applied dynamic programming*, Princeton University Press.
16. W. Yourgrau and S. Mandelstam (1968). *Variational principles in dynamics and quantum theory* (3rd edition), pp. 162–80. Pitman.

17. P. S. Laplace (reprinted 1920). *Essai philosophique sur les probabilités*, p. 7, E. Chiron, Paris.
18. Joseph Needham (1927). *Man a machine*, p. 93, Kegan Paul.
19. H. H. Rosenbrock and P. D. McMorran (1971). Good, bad, or optimal?, *Trans. IEEE*, vol. AC-16, pp. 552–4.
20. H. H. Rosenbrock (1972). The use of computers for designing control systems, *Measurement and Control*, vol. 5, pp. 409–12; (1977). The future of control, Automatica, vol. 13, pp. 389–92; (1988). Engineering as an art, *AI and Society*, vol. 2, pp. 315–20.
21. Hubert Dreyfus (1972). *What computers can't do*, Harper and Row.

3 Hamilton's principle

Purpose in natural systems

1. *Some early history*

A purpose generates a policy, which is the causal relationship that accomplishes the purpose. Our universal practice is to explain nature by causal relations. Can these be derived from a purpose?

The earliest known demonstration that this can, at least sometimes, be done goes back to the second century AD. It was already known at that time that when light is reflected by a mirror, the angle of reflection (Fig. 3.1) is equal to the angle of incidence. Hero showed[1] that this is the policy arising from a purpose: given the initial point A and the final point B on a ray of light reflected from a mirror, let the ray follow the shortest possible path from A to B. As light travels in air at a constant speed, this is the same as saying that light travels from A to B in the shortest time. We therefore have the kind of problem illustrated by the shortest-time-to-climb of an aircraft, which was described in the previous chapter.

Here we are shamelessly translating Hero's result into modern terms, with some violence to the historical facts. Hero knew nothing of schedules or policies, and he believed that we see by rays proceeding from, and not to, our eyes. Nevertheless, the substance of his result is equivalent to the statement we have given.

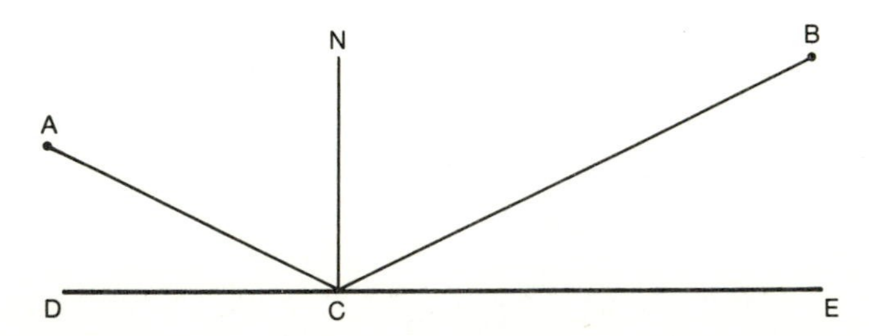

Fig. 3.1 The ray of light from A to B is reflected from the mirror DE at C, and CN is normal to the mirror. The angle of reflection BCN is equal to the angle of incidence ACN, while the plane containing ACB is perpendicular to the mirror.

The proof is nowadays a matter of elementary mathematics, but we do not need even that to convince us of its truth. Represent the ray of light by a thread (Fig. 3.2) which is tied to an attachment at B and led through a small frictionless ring at A. The thread is also led through another (frictionless) ring C on a small floating disc. This second ring will move in a plane, which represents the mirror.

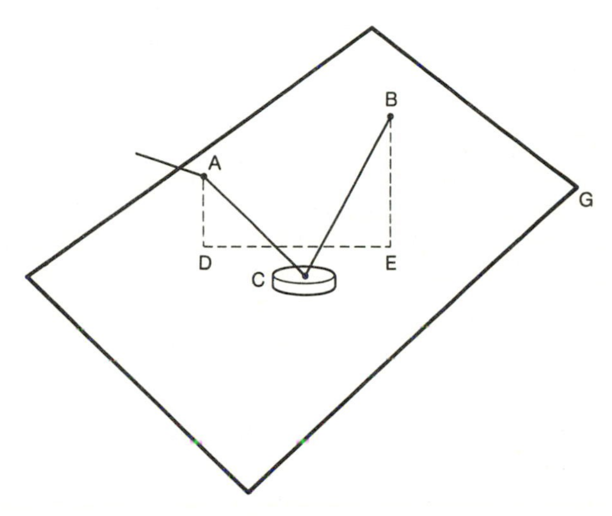

Fig. 3.2 A thread is fixed at B and passes through frictionless rings at A and at C, the latter being mounted on a float so that it moves in a plane G representing the mirror. AD and BE are perpendicular to G. If we pull gently on the thread beyond the ring at A, the float will move until it lies on the line DE. Then the plane ADEB will appear as in Fig. 1, and the ring C on the float will come to rest when the forces exerted by the thread towards D and E are equal. This requires the angles BCE and ACD to be equal. Thus the requirement that the path of the light ray from A to B via the mirror shall be as short as possible gives rise to the condition described in Fig. 3.1, including the fact that AC and CB are straight lines.

Now if we pull gently on the end of the thread beyond the ring at A, the disc will move until the length of thread from A to B is as short as possible. If AD and BE are perpendicular to the plane representing the mirror, the ring on the disc will lie on the line DE. Otherwise, by moving the disc along the line perpendicular to DE until the ring is on DE, we can shorten the thread.

Also, in the plane containing the thread (Fig. 3.1), the force tending to move the disc towards D must be equal to the force tending to move it towards E. The two forces arise from the tension in the thread, which is equal on both sides of the disc because we are neglecting friction at the ring, and these forces can only be equal if the thread makes equal angles to the line DE on both sides of the ring. It will be seen that we have derived not only the usual law of reflection, but also the fact that light rays follow a straight path in air.

In this argument, in order to avoid mathematics, we have appealed to notions of 'force', but the forces belong to our analogy, with its thread and float and frictionless rings. They have nothing to do with the path of light nor with Hero's explanation of it, and the discussion could have been carried out entirely in geometric terms. All we are doing is to show that the requirement that the path from A to B via the mirror should be as short as possible demands that the angle of reflection shall be equal to the angle of incidence. This can easily be proved by elementary geometry.[1]

Hero's result illustrates very well the opposition between explanations in terms of purpose and of causality, and the objections which have been made to the former. It is asked, for example, how light emitted from A can decide among all paths which it might follow to B, which will be the shortest; and how it can then follow this selected path. This is to talk in terms of the schedule, which determines the velocity at each future time before the motion begins.

The policy which accomplishes the purpose is to make the angle of reflection equal to the angle of incidence. This coincides with the ordinary causal explanation, and if we believe that causal explanations are in some sense the 'true' way of explaining nature, then the purpose from which it was derived seems no more than a pointless curiosity.

To anticipate what will be said later, so that misunderstandings do not arise in the meantime, the view which will be presented is as follows. Nature knows nothing of explanations in terms of causality or of purpose: these are human constructs. A ray of light goes where it does, and no-one supposes that it has some way of measuring the angles of incidence and reflection, and ensuring that they are equal. No more need we believe that the ray appeals to the purpose of following the shortest path. Both explanations meet a human need to predict and control, and to that extent to 'understand'.

Over a very wide range of phenomena, and quite possibly all physical phenomena, explanations in terms of purpose, and explanations in terms of cause and effect, are scientifically equivalent. Any evidence which confirms one will confirm the other; any evidence which contradicts one will contradict the other. So for scientific purposes one is as good as the other. But in determining the way we act, the two explanations may be very different. To live in a world from which purpose has been banished is to become one kind of person; to live in a world which acknowledges the existence of purpose is to become a different person.

This theme will be developed later. Meanwhile there is one point we have omitted, which is needed if Hero's explanation is to be equivalent to the usual causal one. It is easy to show that the path of shortest length requires equality of the angles of incidence and reflection. Does the equality of these angles ensure that the path is the shortest possible? The investigation is not difficult, and the result is negative: sometimes the path is the shortest, sometimes it is not.

What is true in all cases is that small changes in the actual path of the ray produce changes in its length which are vanishingly small. That is, as we let the deviations from the actual path become smaller, the changes in the path length get smaller more rapidly than the deviations; a loose but convenient way of describing this is that 'small changes in the path leave its length unchanged'.

This is a property of paths which have minimum length, but also of those which have maximum length, and of some having a length which is neither a maximum nor a minimum (Fig. 3.3). All paths having this property are said to have a length which is 'stationary', and to satisfy a 'variational principle' rather than a minimum principle. The distinction is essential in an accurate treatment, but for informal discussion it is more convenient to talk about a minimum. We shall usually do so, choosing our examples to correspond.

Accordingly, Hero's principle has to be modified: a ray of light reflected by a mirror has a length which is stationary. With this change, the purpose (stationarity) implies the causal relation (equality of angles of incidence and reflection) and vice versa. Sometimes, as in the example of Fig. 1, the length is not only stationary, but is a minimum.

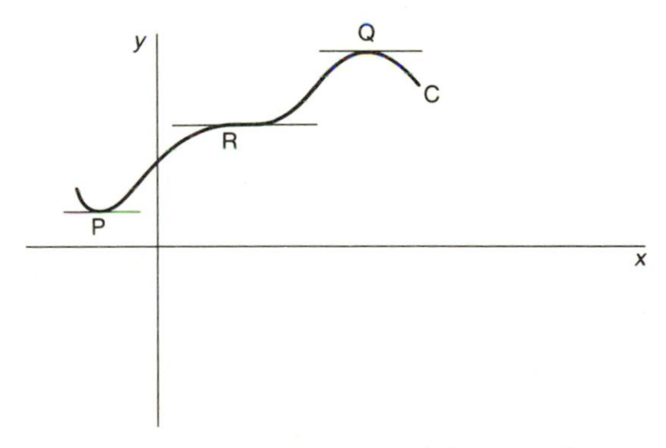

Fig. 3.3 The value of y for any x is shown by the graph C. At P there is a minimum, and the tangent to the graph is horizontal. If x moves slightly away from the point where the tangent is horizontal, the change in y is vanishingly small. The last statement is also true at Q, which is a maximum, and at R which is neither a minimum nor a maximum. P, Q and R are all stationary points.

In Hero's variational principle, y will represent the length of the ray, while x represents the size of a perturbation (of any shape) of its actual path. P, Q and R could all represent the situation for rays satisfying the variational principle.

The purpose, that is stationarity, can be used directly for the solution of problems, just as we normally use causal relations, and without first deriving the policy from the purpose. For example, a well-known way of drawing an ellipse is to put a loop of thread over two pins A and B (Fig. 3.4) and run a pencil C around while keeping the loop tight. This means that the path ACB has constant length, so that it has the stationary property defining the path of a light ray: moving the point C does not change the path length.

If we rotate the ellipse around the line AB, it traces out an 'ellipsoid of revolution'. A mirror having this shape will ensure that every light ray emitted from A, in whatever direction, will be reflected to pass through B. As one of the pins is moved very far away, the ellipsoid becomes a paraboloid: a portion of this will make the reflector for a car headlamp, producing a parallel beam of light if the source (the filament of the bulb) is small enough. Exactly the same conclusion can be drawn from the policy (equality of angles of incidence and reflection) though with somewhat more work.

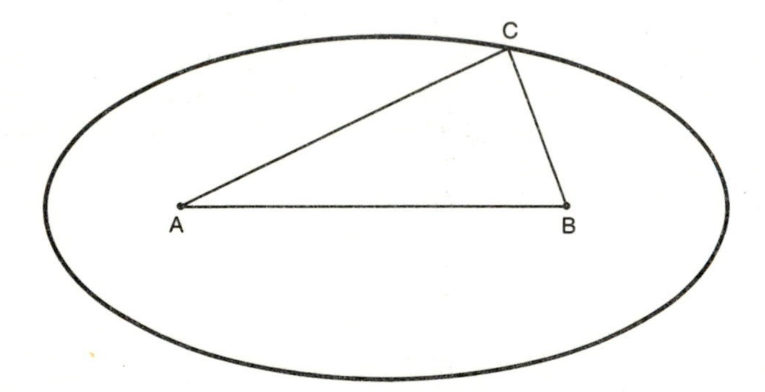

Fig. 3.4 A loop of thread passes round the pins AB, and is held taut by a pencil at C. As the pencil is moved it traces out an ellipse. Since we know that the light rays in air are straight (this readily follows from the variational principle) we need only consider perturbations of ACB which move C along the ellipse. It is clear that all perturbations of this type leave the path length ACB unchanged, wherever C may be on the ellipse. Hence all rays ACB satisfy the variational principle, and can be actual paths for a ray. The argument is readily extended to the 'ellipsoid of revolution' obtained by rotating the ellipse around the line AB.

2. *Further development*

Only in the seventeenth century did the development of mathematics in Europe first equal, and then surpass, the level achieved in antiquity. In that century Fermat and Huyghens gave results[2] which went considerably beyond Hero's. Applying the principle of least time to situations where the speed of light was varying, they were able to deal with refraction, where a ray passes from air into glass or water, and to deduce Snell's law (Fig. 3.5). Generalizing further to a transparent medium in which the velocity of light changes continually, their results define curved paths which will be followed by the rays of light. The mathematical treatment of these approaches closely to that of the minimum-time problem for the climb of an aircraft.

The extension of variational principles from light to the motion of material bodies was proposed by Maupertuis[3] in 1744. His account is not well developed mathematically, and his chief aim is a theological one. He believed that he could demonstrate a minimum principle, rather than a variational principle demanding only a stationary value. He also believed that this was the only description of the laws of motion which was consistent with Christian theology. 'Hence', he says, 'I shall not look for these laws in mechanics, but in the wisdom of the supreme Being.'[4] This mingling of science with theology, threatening the uneasy separation which had been achieved in the previous century, has clouded discussion ever since.

Two major mathematicians, Euler and Lagrange, put the ideas of Maupertuis into a more rigorous and acceptable form as a variational principle. It was, however, little used, and Poisson in 1837 could describe it as 'only a useless rule'.[5] It was with the work of Hamilton and Jacobi in the nineteenth century that the principle took its final form; it is referred to as 'Hamilton's principle'. The form of this principle bears a close resemblance to the purpose which we assumed to underlie the control of our boat in

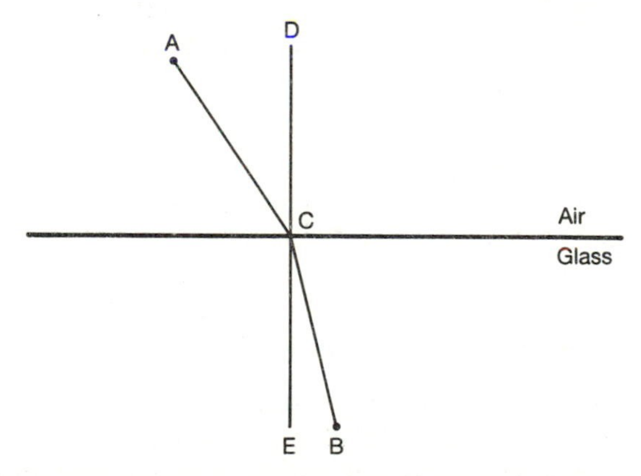

Fig. 3.5 ACB is a ray of light which passes at C from air into glass. The angle of incidence is ACD, and the angle of refraction is ECB. If v_a is the velocity of light in air and v_g is its velocity in glass, Snell's law can be written

$$\frac{\sin \text{ACD}}{\sin \text{ECB}}=\frac{v_a}{v_g}$$

Chapter 2. No surprise need be felt, because the control of the boat was devised so that it could act as an analogy to Hamilton's principle.

If we state this principle for a body such as a stone moving under the action of gravity, it has the following form. The rate at which fuel is consumed by the boat is replaced by something called the Lagrangian of the stone. This is equal to the kinetic energy of the stone less its potential energy, both of which are defined below. The total fuel used corresponds to what is called the 'action': to avoid confusion with the ordinary meaning of the word, 'action' in this sense will always have the quotation marks. For the actual motion the 'action' is stationary: that is, small perturbations of the actual path of the body leave it unchanged. The stationary value of the 'action' is the OCF, the optimal cost function. In many cases the 'action' is not only stationary but also a minimum, and in our discussion we shall assume that this is true.

The promised definitions are as follows:

i. The *kinetic energy* is the energy which the stone has by virtue of its motion. The kinetic energy of a car is the energy which has to be put into it to accelerate it, and which destroys a car which is crashed into a solid obstacle at speed. Kinetic energy is proportional to mass, and proportional to the square of the speed (provided that this is small compared with the speed of light). The fuel consumption of our boat was assumed to depend directly on its displacement (which, appropriately defined, is its mass) and on the square of its speed. Hence the rate at which fuel is used to propel the boat is a good analogy to kinetic energy for the stone. It is no more than an analogy, of course: the stone does not *consume* kinetic energy.

ii. *Potential energy*, for a body subject to gravity, is the energy which it has by virtue of the height through which it has been raised. It is proportional to the mass of the body, and proportional to the height through which it has been raised (provided that this is small compared with the radius of the earth). When a car climbs a hill, potential energy is stored in it. If the car is driven over a cliff, the potential energy is converted into kinetic energy as it falls, and this energy will destroy it when it meets the ground. We have not so far introduced anything which will represent potential energy for our boat, but we shall now do so.

In Chapter 2, we saw that in still water our boat sailed at a constant speed in a constant direction in order to fulfil the purpose we assigned. A tidal flow can cause it to depart from a straight course and a constant speed, and is an entirely natural thing to introduce. It is a good analogy for certain kinds of force acting upon our stone, but not for the gravitational force. To represent this we have to make a somewhat artificial assumption.

Suppose that our boat is a passenger liner, sailing in the latitudes between 20° and 30°N. Temperatures there are high, and the ship is equipped with air conditioning which consumes a certain amount of fuel. The rate at which fuel is used for this purpose depends directly on the displacement of the ship, and increases as the external temperature increases. We suppose that this increase is proportional to the decrease in latitude: one degree nearer the equator increases the rate at which fuel is used by the same amount wherever we are within this general area. We neglect the curvature of the earth and its rotation, which have only a small effect, and we suppose there is no tide.

Now suppose the ship is to sail 100 km east from A along the twentieth degree of latitude to B, in five hours (Fig. 3.6). In doing so, it will use some fuel for propulsion and some for air conditioning. Neglecting the latter, the best course would be to sail due east at 20 km per hour, which would demand a certain total consumption of fuel for propulsion. Adding the fuel needed for air conditioning during 5 hours at latitude 20°, we obtain the total fuel consumed during the 5 hours, for all purposes, when sailing due east at 20 km/h.

It is evident that if the straight course is abandoned for one that deviates to the north, somewhat more fuel will be needed for propulsion, and less for air conditioning. We can guess that a small deviation of this sort may make the gain from the latter effect greater than the loss from the former, and that a best (minimum total fuel) course will exist which lies to the north of the previous course. A simple calculation shows that this is so. The optimal policy is to maintain the component of speed towards the east constant at 20 km/h. To this we add a component of speed which is initially to the north, but which decreases steadily to zero and then increases steadily to the south.

The rate of decrease or increase (both being represented by an acceleration to the south) is constant during the voyage, and we find that the optimal path is a parabola (Fig. 3.6). The total length of the path is somewhat increased, and so also therefore is the speed, so that more fuel is used for propulsion in the given time, but less is used for air conditioning.

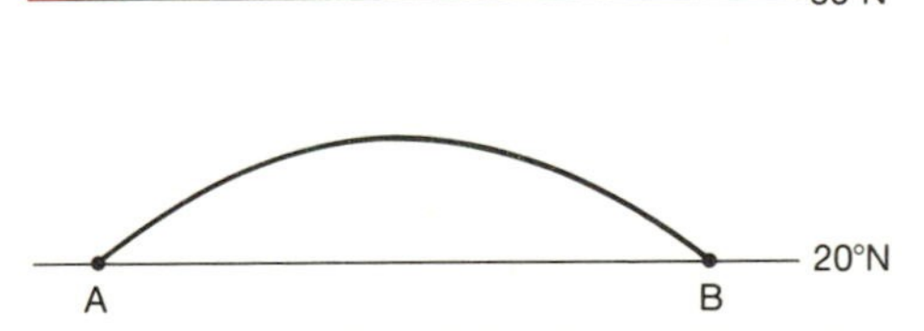

Fig. 3.6 A ship sails from A to the point B 100 km due east on the latitude 20°N. We neglect the curvature and rotation of the earth and assume that wind and water are still. Fuel is used for propulsion, proportional to the displacement of the ship and the square of its speed. Fuel is also used for air conditioning, and this decreases uniformly for every degree further north. The total fuel used will be least for a curved path (actually a parabola) deviating to the north by an amount which increases as the time allowed for the voyage increases.

The resulting path has the optimal property: any deviation which we make from the optimal course or speed at any point will increase the total amount of fuel used.

It is easy to see that the deviation from a straight path will be small when the time allowed for the voyage is small, because then the amount of fuel used for propulsion is large, and for air conditioning is small. Gains in the latter from deviations to the north will only pay for very small increases in the former. As the time allowed for the voyage increases, the balance will change, and larger deviations to the north will become profitable.

We have now created an analogy for our stone moving under the effect of gravity. The rate at which fuel is used for propulsion corresponds to kinetic energy. The rate at which fuel is used for air conditioning corresponds to potential energy. The total fuel used corresponds to the 'action', and the stone moves so as to make this 'action' a minimum (more generally, stationary, but also a minimum in the case we are discussing).

The course of the ship represents the path of the stone, because we have arranged for the mathematical description of both to be the same. To be sure, there is an immense difference of scale, but the numbers representing each can be made the same by choosing say kilometres in measuring the progress of the ship, and decimetres for the stone; and similarly for other variables.

3. *Motion of the stone*

As in Chapter 2, we can describe the motion of our hypothetical ship in three ways.

i. The captain can sail a course with the purpose of going from A to B in a given time with the minimum consumption of fuel.

ii. An engineer can design an automatic pilot embodying a schedule to implement the purpose.

iii. The engineer can embody in the autopilot a policy which implements the purpose.

In all cases the externally observed behaviour will be the same (remembering that we have excluded random disturbances) but our explanations will be different.

i. We should say the captain sailed the ship in such a way as to fulfil his purpose. It pays him to divert towards the north because this reduces his total fuel consumption.

ii. The ship following a schedule pursues a course which is predetermined.

iii. The ship following a policy obeys causal laws.

In the same way we can have three different explanations for the observed behaviour of the stone.

i. It fulfils the purpose defined by Hamilton's principle. It has to go from A to B in a given time (Fig. 3.7) while minimizing the 'action', which corresponds to the total fuel used by the ship. The Lagrangian, which corresponds to the rate at which fuel is used, is equal to the kinetic energy (increasing with the square of the speed) *less* the potential energy (proportional to height). Notice the subtraction, which comes from the way we define potential energy: if we used the less common 'work function' we should have addition. The result is that the Lagrangian (corresponding to the rate of using fuel) is reduced by increasing the potential energy, that is, by choosing a higher path. The optimal path is a parabola, as for the ship. It should be noted that in saying the stone fulfils a purpose we are not attributing to it either consciousness or will. The stone goes where it does: the purpose is our explanation.

ii. The stone follows a predetermined course. The possibility of this description, corresponding to the schedule, has caused considerable confusion in the past. It has been proposed that the whole course of nature, the behaviour of everything in the world, was uniquely determined at its creation. Once random behaviour is admitted, as it is in quantum theory, the difficulty largely disappears. The detailed future course of events can no longer be predicted.

iii. The stone follows causal laws. These are: that the horizontal component of its velocity shall remain constant, while it starts with a vertical velocity which continually decreases, becomes zero, and is followed by an increasing downward velocity. The vertical acceleration, which is the rate of change of the upward velocity, is constant and negative: as we say, the stone accelerates downwards.

It is usual to explain this downward acceleration, as Newton did, by saying that it is produced by the downward force of gravity acting on the stone. When we use Hamilton's principle, the idea of force is not needed—we did not use it in explaining the motion of the ship, and though we could have introduced a force for that purpose it would have seemed pointless to do so. Nevertheless there is no objection to using the idea of force, and some convenience: its prevalence no doubt springs from our subjective experience of a downward force when we hold a heavy stone.

If we think in terms of forces, the motion of the stone at first seems paradoxical. The gravitational force acts downwards, yet the path which the stone must follow to minimize the 'action' makes an upward curve, and the greater the force, the greater the upward deviation from a straight line. The reason lies in the constraint we have imposed, that the stone must go from A to B in a given time. Then it is clear from the usual way of regarding its motion in terms of forces that the constraint requires it to follow the path already shown in Fig. 3.7.

Though almost all discussions of Hamilton's principle use the constraint we have given, other constraints can be used, and in quantum theory other constraints are forced upon us. One alternative constraint is particularly important, and we can approach it in the following way. Let us start from the solution of the problem in which A and B are specified, together with the time of transit. The OCF for this problem defines the initial velocity at A. Conversely, if we define the initial position A, and the initial velocity v, this defines the OCF, except that we can add any constant number we please to the OCF and it still remains a solution.

The reason for this is that many other points B′ will produce the same v at A, namely all points on the optimal trajectory, Fig. 3.7. For each of these the OCF will have a different value: in terms of our ship, the journeys to these different points will all require different total amounts of fuel. (It will be remembered that the OCF was the least amount of fuel needed to reach the final point from where we are now). The new constraint (A and v) is as effective as the old one (A, B, and T) in avoiding the trivial solution, in which the ship (or the stone) remains always at A.

Fig. 3.7 A stone is thrown from A so as to go to B, at the same elevation. It moves so as to minimize the 'action', and this is achieved by deviating upwards (along a parabolic path) from the straight line AB. The extent of the deviation increases as the specified time of transit from A to B is increased.

If B′ is a point on the optimal trajectory from A to B (which is the actual trajectory) and is reached in a time T' then the portion AB′ of the original path AB′B solves the problem 'go from A to B′ in the time T' while minimizing the "action".'

4. *Recapitulation*

As we have covered a large territory in the development of this discussion it may be helpful to recapitulate the conclusions we have drawn from Hamilton's principle. This specifies an objective, or purpose, and the analogy of our ship provides us with a simple way of appreciating its significance.

When we throw a stone into the air, we can if we wish say that it moves in such a way that it satisfies Hamilton's principle: it fulfils a purpose. The fact that it does so is all that is needed to predict the path which it will follow. We can make this prediction when we know where it starts, where it ends, and the time it takes: in this case we predict the course it follows to satisfy these conditions. Alternatively, we can make the prediction knowing the initial point and the initial velocity (that is, the initial speed and direction). We then predict the course it follows from the initial point. In each case, our prediction gives us the points on the trajectory, together with the velocity and time corresponding to each of them.

The calculation, in principle, can be made by categorizing all possible solutions in some way, comparing the values of the 'action' which they produce, and picking the minimum (more generally, the one giving a stationary value). In practice, the labour involved in doing this makes the procedure impracticable, but this difficulty does not affect the principle.

Alternatively, the purpose defined by Hamilton's principle can be translated into a policy. This comprises the causal relations which will accomplish the purpose, and for a stone moving under the influence of gravity the causal relations are Newton's equations. From these causal relations the path of the stone can be calculated, and it will be the same as was found directly from the purpose.

As another alternative, we can translate the purpose into a schedule which tells us directly where the stone will be at all times after it starts. This procedure is less interesting to us because it does not fit in so well with later developments. In all three cases we obtain the same predicted course for the stone (we are neglecting random disturbances) but the procedure is different in each case. In the first we use the untranslated purpose. In the second we first translate the purpose into a policy. In the third, we translate the purpose into a schedule.

A conventional view is that the causal description, that is the policy, is the 'true' description of nature. If so, the existence of the purpose, and the fact that the causal relations can be obtained from it by generating the policy, are of minor interest. These facts will be used if they are convenient, but only because they are equivalent to the 'true' description in terms of cause and effect.

The view adopted here, on the other hand, is that the purpose, and the policy into which we can translate it, are scientifically equivalent. Any evidence which supports one supports the other. Any evidence which contradicts one contradicts the other. Hence, within the scientific framework, we are not entitled to prefer one to the other: to do so would be to claim an a priori knowledge about nature which we do not have.

It is this topic, of equivalent theories and our attitude to them, which will be discussed in the following chapter. Before doing so we consider some further questions concerning Hamilton's principle.

5. *Scope of Hamilton's principle*

For simplicity, we have introduced Hamilton's principle as an explanation for the motion of a stone under the influence of gravity. Its scope in fact is very wide, and encompasses most of what is called 'classical mechanics', which is essentially physics as it developed up to the third quarter of the nineteenth century. It is concerned with things that are not too small, and do not travel too fast. The second kind of exception, over speed, was the first to arise, and it led to the theory of relativity. The first gave rise to quantum theory, and both these developments can be dealt with by changes in Hamilton's principle as will be explained later.

The changes which have to be made are extensions, which leave Hamilton's principle practically unchanged under the conditions which were accepted in classical mechanics. In order to deal with speeds approaching the speed of light, we modify the Lagrangian, corresponding to the fuel consumption of our ship. But the modified Lagrangian is almost the same as the classical one for speeds which are small compared with the speed of light. To deal with very small objects such as electrons, we shall add a random disturbance to Hamilton's principle. The random disturbance should also be added for larger objects, but its effect is then so small that we can ignore it.

Over a wide range, therefore, Hamilton's principle holds in the form in which we have described it. The motion of a stone can be derived from it. So can the motion of a satellite outside the earth's atmosphere, or the motion of the earth around the sun. An electrically charged particle, neither too small nor too fast, moves in an electric field according to the principle: its motion in a magnetic field obeys the same principle provided that we use the appropriate Lagrangian. The purpose represented by Hamilton's principle remains constant in all these applications. What changes is the Lagrangian, which in our analogy is the formula defining the rate at which fuel is used.

In the cases mentioned, we are concerned with particles moving under the influence of gravitational or electromagnetic forces. Many of the systems we meet in practice involve other variables such as the flow of heat, or the forces which constrain the wheel of a car to rotate around its bearing. The relation of these to our principle is most easily considered at a later stage, and they will be discussed in Section 10 of Chapter 5 when we look at subordinate purposes.

6. *Causality*

When a distinction was made earlier between efficient causes (which we call simply causes) and final causes (which we call purposes), the former were assumed to be unproblematic. We have a subjective experience of cause and effect: we push a car and it moves, or we throw a stone and it flies through the air.

Yet historically the idea has been the subject of much debate. We shall never call one thing the cause of another which precedes it, so that causality implies non-anticipation. But if one event always precedes, and is always followed by, another, we do not always say that the first caused the second. The flash of lightning is always followed by the rumble of thunder, but we do not say that the light causes the sound. We say rather that both are caused by an electrical discharge from the clouds, and that the light reaches us before the sound.

There are two chief things which lead us to say that a causal relation holds between two events. The first is the existence of a theory in which one is cause and the other effect. The electrical discharge causes the flash of lightning by a mechanism which we can describe. It also causes the sound of thunder by a different mechanism. Therefore we do not say that lightning causes thunder. The full explanation must be quantitative, because a sufficiently powerful flash of light, as from a laser, could very likely cause a noise like thunder; but in nature we believe that the mechanism is different.

A pleasing story illustrates the connection between theory and causality. Its origins are now lost to me, but it concerns a scientist who formed a friendship with an Australian aborigine, and spent many evenings in discussion with him around a small camp fire. Once he was explaining the cause of the tides: how the moon exerts an attraction on the earth, and this attraction causes water to pile up on the side which faces the moon. Then, as the earth rotates under the moon, the piled-up water moves around the earth and produces one of the two tides which occur each day. And how the second tide each day is due to a second piling up of the water, which occurs on the side of the earth facing away from the moon, and moves round the earth in the same way as the other.

At this point the scientist's explanation faltered, because it is not easy without some mathematics to explain how this second bulge in the sea arises. Nevertheless his friend listened politely, and there was a pause before he smiled and said, 'That is very interesting. But of course everyone knows that the tides are really caused by the crabs on the sea-shore. When the moon is in the right place they come out of their holes in the sand, and the water runs away, and that is what causes the tides. And anyone can convince himself that this is true by the evidence of his own eyes, without long explanations. All he need do is to go down to the shore when the tide is out, and he will see thousands of crabs upon the sand'.

For us, the receding tide is the cause, and the appearance of the crabs is the effect. If we were called upon to justify our belief, one way would be to demonstrate that our theoretical explanation was better than the alternative, having greater coherence and explanatory power, and greater scope. But there is also another way, which introduces the second justification which we commonly give for attributing a causal relationship. This is to manipulate the cause, if we can, and see whether or not this alters the effect. We might, for example, propose to kill all the crabs, as the fire-ant was once proposed for extermination,[6] and see whether the tides ceased. If not, we should assert that the crabs did not cause the tide. This second justification for attributing a causal relationship is universal in technology. We accept that moving the rudder is the cause of a change in heading because we use the rudder to steer a boat, and there is a connection here with our subjective experience of purpose.

We talked earlier about an infinite chain of cause and effect: A causes B as its effect, but B causes C causes D, and so on. This is one way in which we regard the world, but subjectively we often see the causal chain in a different way: it originates in the adoption of a purpose; as when we change the heading of a boat by moving the rudder, so initiating a causal chain to fulfil our intention.

The traditional scientific experiment has a similar structure. Let the initial conditions be the same in two cases, we say, and in the one case let us perform a certain action, and in the other case refrain from performing it. Then the difference in outcome in the first case is the effect of the action we took. That is, we are proposing that initially the two experimental systems are at the same point in identical causal chains. In one case we modify the causal chain by our action, and in the other we leave it unchanged.

It is always possible to say that the experimental system which we modified, together with ourself, formed a single infinite causal chain, and that no new causal chain was initiated. But this is not the way we ordinarily think of the matter, as Socrates made clear some two and a half thousand years ago. In Plato's account,[7] Socrates is in prison, voluntarily awaiting death, though his friends have urged him to flee and provided the means to do so. He justifies his resolution to remain and to submit to the judgement of the Athenians, contrasting his decision, based on what he considers right, with the causal explanations of some philosophers of his day.

And it seemed to me it was very much as if one should say that Socrates does with intelligence whatever he does, and then, in trying to give the causes of the particular thing I do, should say first that I am now sitting here because my body is composed of bones and sinews, and the bones are hard and have joints which divide them, and the sinews can be contracted and relaxed and, with the flesh and the skin which contains them all, are laid about the bones; and so, as the bones are hung loose in their ligaments, the sinews, by contracting and relaxing, make me able to bend my limbs now, and that is the cause of my sitting here with my legs bent. Or as if in the same way he should give voice and air and hearing and countless other things of the sort as causes for our talking with each other, and should fail to mention the real causes, which are, that the Athenians decided that it was best to condemn me, and therefore I have decided that it was best for me to sit here and that it is right for me to stay and undergo whatever penalty they order. For, by the Dog, I fancy these bones and sinews of mine would have been in Megara or Boeotia long ago, carried thither by an opinion of what was best, if I did not think it was better and nobler to endure any penalty the city may inflict rather than to escape and run away. But it is most absurd to call things of that sort causes. If anyone were to say that I could not have done what I thought proper if I had not bones and sinews and other things that I have, he would be right. But to say that those things are the cause of my doing what I do, and that I act with intelligence, but not from choice of what is best, would be an extremely careless way of talking. Whoever talks in that way is unable to make a distinction and to see that in reality a cause is one thing, and the thing without which the cause could never be a cause is quite another thing.

Socrates here describes very clearly the process of adopting a purpose, and of translating the purpose into those causal relations which will accomplish it. This is also the procedure adopted by the control engineer. He is given the purpose, and he translates it into the appropriate causal relations. But then, instead of carrying out the necessary actions himself, or arranging for them to be carried out by someone else, he builds a machine to perform them.

The statement that 'man is a machine'[8] is intended to contradict both of these references to purpose. It envisages an explanation of Socrates' behaviour, not simply by bone and muscle, but also by mental events due to past experience and hereditary tendency, without any reference to purpose. Similarly, it would deny that a human purpose existed which could serve as an origin for the causal relations built into the control engineer's machine: the whole chain of events would be seen as causal, and without purpose.

In each case there is a conflict with our ordinary way of thinking and talking. Our main argument, however, does not rely on this conflict, but rather on the equivalence which can be drawn between purpose and causality. We drop, for the sake of our main argument, any reference to 'adopting' or 'forming' a purpose, which raises the problem of free will. Then, as we have shown, Hamilton's principle defines a purpose which is equivalent to the causal description of a wide range of classical systems. It can be extended to relativistic systems, as will be shown later, and also to quantum-mechanical systems. We therefore turn to the implications of equivalence for scientific theory.

References

1. W. Yourgrau and S. Mandelstam (1968). *Variational principles in dynamics and quantum theory* (3rd edition), pp. 5, 11, 12, Pitman.
2. Reference 1, pp. 11–18.
3. Reference 1, pp. 19–23.

4. Maupertuis (reprinted 1984). *Essai de cosmologie*, etc., p. 35, J. Vrin (Present author's translation).
5. Reference 1, p. 32.
6. Rachel Carson (1962). *Silent spring*, Penguin edition 1965, pp. 147–57 (Penguin Books).
7. Plato, translated by H. N. Fowler, vol. 1, *Phaedo*, pp. 339–40, Heinemann.
9. Joseph Needham (1927). *Man a machine*, p. 93, Kegan Paul.

4 Equivalence

The underdetermination of theory

1. *The foundations of science*

One of the implicit assumptions of the scientific method is that our theoretical explanations of nature are fully determined. There may at any time be competing theories, particularly when an accepted theory is being questioned. But it is always assumed that it will be possible, with enough effort, to bring forward facts which will eliminate one contender leaving, at least for a time, a unique and generally accepted theory.

The history of modern science certainly supports this view. It has been very rare for two or more alternatives to coexist on fairly equal terms for any long period. Nevertheless, the assumption has been challenged, particularly in the last thirty years: the opposing doctrine is known as the 'underdetermination of theory'. It is supported from two sides, from sociology and from philosophy, on different grounds.

In the sociology of science, our attempt to understand and explain the world we live in is regarded as a part of the whole experience that we have of living, and of living necessarily in a society. Our beliefs are seen as one part of this experience, reacting upon and reacted upon by the rest of our experience. From this point of view, the explanation of the tides by the theory of gravitation, or their explanation described in Chapter 3 by the activity of crabs upon the seashore, are of equal status as objects of study.

Whether they are of equal status as embodiments of truth (however defined), or in power of conviction, or as guides to action, can then come to be treated as questions which themselves depend for their answers largely upon social determinants. Carried to an extreme, this can result in the view that one theory of the world is as good as any other, given the appropriate social context.

A young man of my acquaintance is accustomed to make this point in his seminars by maintaining that Newton's laws of motion do not express anything about the natural world, but rather something about the beliefs of certain kinds of society. He is then, predictably, invited by scientists in his audience to jump out of a third-floor window. His declining to do so is taken on the one side as a conclusive disproof of his position, and on the other as logically beside the point. And if he had accepted the invitation to jump out of the window, the natural consequences would again have been taken by the scientifically inclined as a validation of their view.

Such extreme views are not common, but more moderate accounts may still have a high degree of relativism:

> [The] message of the work . . . is relativistic because it suggests that belief systems cannot be objectively ranked in terms of their proximity to reality or their rationality. This is not to say that practical choices between belief systems are at all difficult to make, or that I myself am not clear as to my own preferences. It is merely that the extent to which such preferences can be justified, or made compelling to others, is limited.[1]

This is clearly in opposition to the view of science held by most working scientists. It makes the choice between gravitation and the crabs upon the seashore a matter of preference, determined at least partly by the social matrix within which we exist. The scientist, on the other hand, sees it as a choice between truth (even if only a provisional truth subject to refinement) and demonstrable error.

Because science is a powerful and successful and well-funded activity, it is largely immune to sociologically based questioning of its claims. Many scientists will be aware of the questioning, and interested in it. But faced with problems in the actual practice of their work, they will approach them with the currently accepted attitudes of science. Any other approach would probably inhibit their success, and would certainly prejudice the funding of their research or the acceptance of their results.

Similar comments, though with somewhat less force, can be made about questioning which arises from the philosophy of science. This limits its attention more or less closely to the activities of the scientific community, rather than seeing this as only a part of a larger social whole. It is concerned with the way in which scientists ought to behave in their scientific activity, how their practice conforms to or differs from this norm, and how their results can be validated or criticized.

Earlier work tended to accept the validity of the scientific method and attempted to define a methodology by which it operated, or ideally should operate. This usually involved the generalization from experiment by way of induction, but the difficulties encountered in the attempt to illustrate and justify this procedure have diminished its influence.

An alternative and influential account given by Popper[2] starts with the formulation by the scientist of a theory. This is no doubt based in some way on observed facts, but it is regarded as given, without inquiry into its source. To be accepted as scientific the theory must be capable of being refuted by observed facts. Scientific effort is then devoted to refuting the theory, and if it survives this effort it is adopted, at least provisionally, as valid. Scientific truth thus comes to consist in all those theories which have not yet been refuted. This is probably the most widely accepted account among the 'hard' sciences such as physics or chemistry.

All such methodological proposals, however, have been strongly questioned by recent philosophers of science,[3] who have shown that what actually happens in science often bears little relation to any of the systematic accounts. As a part of this questioning it is often asserted that any set of facts is consistent with more than one theoretical structure which can be erected for their explanation.[4] The theories which are consistent with some particular set of data are not, in the usual account, identical in their predictions, and more extended data may therefore eliminate some, while leaving others still in contention.

All working scientists are probably aware in outline of these developments, and they have had an effect on the perception, by scientists, of science. But in the actual, day-to-day practice of science they have again had rather little influence. They come from outside (though close to) the active scientific community, which carries with it its own practices and standards of judgement.

Now the main line of argument which we wish to pursue here is the contention that science incorporates values which are not a part of science, but which profoundly affect the way we view the world, and the way we behave. In particular, science eliminates purpose. It would be possible to develop this argument on a basis of the sociology of science, or of the philosophy of science, but no argument on either of these bases would be likely to carry conviction where that is most needed.

Accordingly, the approach we shall use will be different. We start by accepting the scientific outlook in its conventional form. For convenience, this will be the form enunciated by Popper, but little change would be needed in the argument if we adopted an alternative form. Then we shall show how alternative theories can be scientifically equivalent in an exact sense. In terms of the Popperian account, the alternative theories will be refuted by precisely the same observations, and will fail to be refuted by precisely the same observations. Any choice which is made between them must therefore be made on grounds which are outside science, because it can never be supported by any empirical evidence.

The grounds of any such choice may well be reasonable and defensible—for example we shall generally prefer a simple to a more complicated theory. But the choice will give no basis for conclusions about 'the true nature of the world': it will merely reflect our preferences. Indeed, we shall maintain that the individual theories which are equivalent are not themselves the appropriate objects which we should regard as a 'scientific description of the phenomena', and against which the process of attempted refutation should be employed. The appropriate object for this purpose is the 'equivalence class' to which they all belong. To explain this, we need some ideas from the mathematical theory of equivalence.

2. *Mathematical equivalence*

An explanation of the mathematical theory of equivalence can fortunately be given without the use of any formal mathematics. The theory starts by supposing that we have a set of objects, and for the sake of example we suppose that these are simple closed figures bounded by straight lines, Fig. 4.1. No figures exist with 1 or 2 sides, but there are infinitely many with 3 sides, or 4 sides, or any larger number. The 3-sided figures, for example, are all the possible triangles of all possible sizes and shapes.

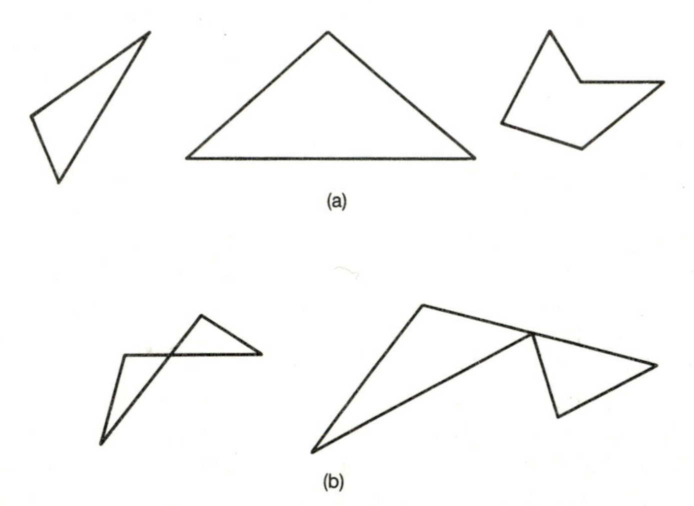

Fig. 4.1 We consider a set of objects consisting of all simple plane figures bounded by straight lines. 'Simple' means that the figures enclose an area which is 'all in one piece'. Some examples are shown in (a), while in (b) are shown two examples which are excluded because they are not 'simple'.

Now what we mean by 'equivalence' is open to choice, though the choice cannot be arbitrary. Name any three of our figures A, B, and C. Then any definition which we choose for equivalence must satisfy three conditions:

i. A is equivalent to itself.

ii. If A is equivalent to B, then B is equivalent to A.

iii. If A is equivalent to B, and B is equivalent to C, then A is equivalent to C.

Here, A, B, and C can stand for any arbitrarily chosen members of our set of straight-sided figures. The three conditions obviously correspond to what we imply in ordinary speech when we say things are equivalent.

One equivalence relation we could choose among our figures would be 'having equal area'. Then given a single figure A, say a triangle, we could look for other figures—triangles of different shape, quadrilaterals, etc.—all having the same area as A. This definition of equivalence obviously satisfies the three conditions.

An alternative choice of equivalence relation is this: two figures are equivalent if they have the same number of sides. Thus all triangles are now equivalent, all quadrilaterals are equivalent, and so on for 5-sided figures, 6-sided figures etc; but no triangle is equivalent to a quadrilateral or to any figure having 5 sides, 6 sides, etc. It is obvious that the three conditions are satisfied again with this quite different relation of equivalence.

In practice, we choose our equivalence relation according to what we are interested in: for example if we are interested in what all figures with the same number of sides have in common, we regard them as equivalent. By doing this, we divide our original set of figures into subsets which we call 'equivalence classes': triangles, quadrilaterals, 5-sided figures, etc. Then we ask what else the members of an equivalence class have in common, besides the defining relation. One answer, in our example, is that they have the same number of angles. If we had chosen the other kind of equivalence, by area, we should have divided our set of figures into a different collection of equivalence classes, all the members of any one equivalence class having the same area. Then we could have looked again for any other property which all members of an equivalence class had in common.

There is a straightforward general procedure for finding out whether a given figure A belongs to a given equivalence class. We choose any member of the equivalence class, say X, and see whether it is equivalent to A. If so, A belongs to the equivalence class, and is equivalent to every one of its members. If not, A does not belong to the equivalence class. In order to carry out this procedure conveniently, we can nominate one member from each equivalence class to represent it, and to act as X in the procedure that has been described.

All of this may seem obvious, and somewhat pedantic; but it has the merit of being general, so that we can apply it to any sort of equivalence whatever. Let us say then, that two scientific theories are equivalent if every fact which refutes one also refutes the other, while every fact which fails to refute one also fails to refute the other. This satisfies the three conditions for equivalence. Then if we start with the set of all scientific theories, our equivalence relation divides this set into distinct equivalence classes: all the members of any one class are scientifically equivalent in the sense we have defined.

The 'set of all scientific theories' may cause us some difficulty. But we can equally well start with 'the set of all scientific theories which have so far been put forward'. Or we can start with a particular set of theories which we list. In all cases we generate equivalence classes, though in the last case we might find that each class had only one member: no two theories in our list might be equivalent. Difficulties of this sort will disappear when we take particular cases.

Following the practice of mathematicians, we can think of the equivalence classes as new objects. 'The triangles' are one example, and any one of our equivalence classes of theories is another. We regard all the members of the class as being the same for our particular purposes, so we can talk of the class as a whole. We can also nominate any member of an equivalence class to represent it: a representative triangle, or a representative theory. We must be careful, however, not to confuse the representative with the class it represents. This will all become clearer if we consider an example.

3. *The world turned inside-out*

What we shall do is to invent an alternative to our usual picture of the world, and we shall do this in a way which makes it scientifically equivalent. Let us call our usual picture the real world, and as we must refer to it many times we abbreviate this to RW. Then our alternative is the inside-out world, or IOW, and we construct this using the mathematical device of inversion. (Bellman has humorously defined a mathematical device as a trick which works at least twice.)

To set up an inversion, we choose a point, any point, and a sphere centred on it—a sphere of any radius we like. For the point we shall take the centre of the earth, which we label O, and for the radius of the sphere we shall take the average radius of the earth, which is about 6400 km and which we call *R*. Then any point A in RW becomes, after inversion, a point B in IOW which is defined in the following way (Fig. 4.2). In RW draw a straight line from the centre of the earth through A. Then B is on the same straight line from O as A, and the distance OB is as many times less than the radius R of the sphere as OA is greater.

That is, if OA is 2R, so that A is a point about 6400 km above the surface of the earth, then OB is $\frac{1}{2}$ R, and B is a point inside the hollow sphere which represents the earth in IOW. On the other hand, if OA is $\frac{1}{2}$ R, then OB is 2R: points outside the sphere transform into points inside it, while points inside the sphere transform into points outside it. Points on the sphere obviously transform into themselves.

In RW, the earth is slightly flattened at the poles, so that in IOW it is slightly raised there, and the hollow IOW earth is egg-shaped. Let us neglect this fact, which does not affect our argument and is easily accounted for

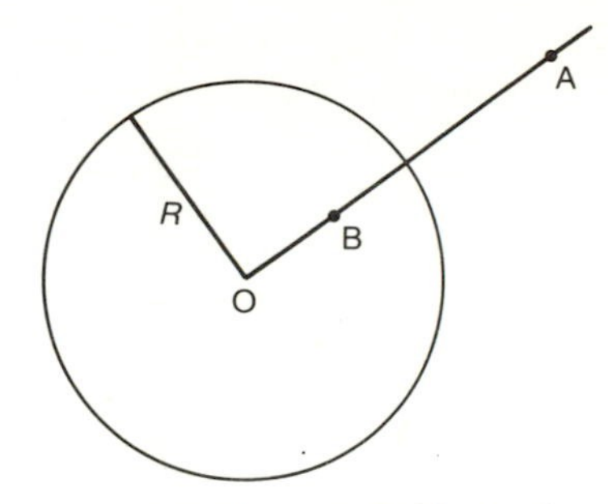

Fig. 4.2 The process of inversion. O is the centre of the Earth in RW, and R is the average radius of the Earth. A is a point in RW, and B is the corresponding point in IOW. If OA is $2R$, then OB is $\frac{1}{2}R$, and so on. Points outside the sphere with radius R in RW become points inside this sphere in IOW, and vice versa. Points on the sphere, which represents approximately the surface of the Earth in RW, are unchanged by the inversion. They become points on the surface of the Earth, which is now a hollow sphere, in IOW.

later, if we wish. Then the mountains in RW become inward projections of the hollow sphere of the earth in IOW, and the depths of the sea in RW are pressed outward in the hollow IOW earth. In IOW, we walk on the inside of the hollow sphere which is the earth, with our heads towards its centre (Fig. 4.3).

Now it is an elementary exercise to show that inversion transforms spheres into spheres (with one apparent exception which we shall explain shortly). We have already had one example of this: the sphere of radius R transforms into itself. In a similar way, but more surprisingly, if we represent the sun by a sphere in RW, it becomes a sphere in IOW.

In RW, the sun can be regarded as spherical, with diameter about 1 400 000 km, situated at a distance of about 150 million km from the earth. In IOW, the sun becomes a small sphere inside the hollow earth, having a diameter of about 2.5 m and placed at a distance of about 270 m from the centre of the earth. The moon in IOW is a larger sphere, about 960 m in diameter and moving in an orbit about 106 km in diameter around the centre of the earth. The fixed stars are minute bodies, all of them being contained in a sphere at the centre of the IOW earth, with a diameter of 2.2 mm (Fig. 4.4).

The exception that was mentioned earlier is that certain spheres, when they are inverted, do not transform into spheres but into flat planes. Specifically, any sphere which has the centre of the earth on its surface transforms into a plane. Conversely, any plane transforms into a sphere with the centre of the earth as one point on its surface (Fig. 4.5).

This exception is easy to understand if we consider larger and larger spheres, say in RW. The point on these spheres which is furthest from the centre of the earth in RW transforms into a point near the centre of the

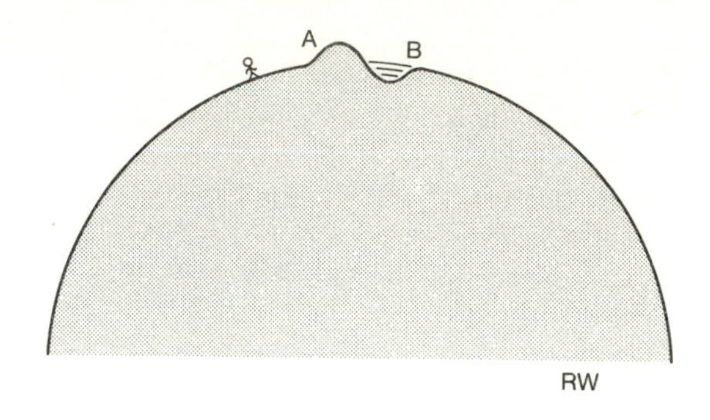

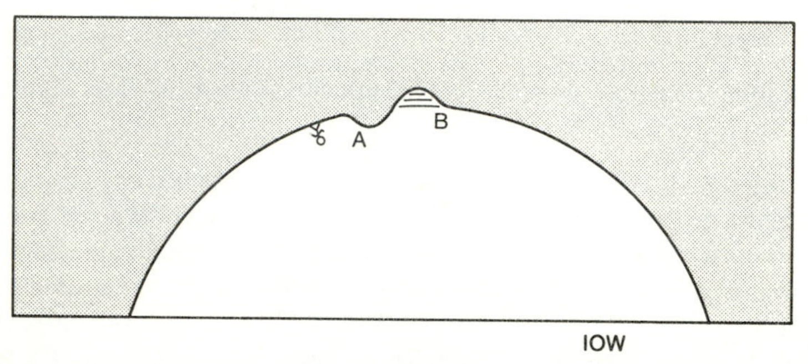

Fig. 4.3 In RW we inhabit the outside of a solid sphere. A represents a mountain and B a sea. In IOW we live on the inside of a hollow sphere which extends outwards indefinitely under our feet. Again A shows a mountain and B a sea.

earth in IOW. As the RW spheres get larger, the IOW spheres into which they transform pass nearer and nearer to the centre of the IOW earth. We can regard a plane as the limiting case of an infinitely large sphere, which transforms into a sphere having the centre of the IOW earth on its surface. In conformity with this, we say that the 'point at infinity' in RW transforms into the centre of the IOW earth.

Everything we have said about spheres and planes applies with appropriate changes to circles and straight lines. Circles, after inversion, become circles; with the exception that circles which pass through the centre of the RW earth invert into straight lines, and so on. This follows at once from the behaviour of spheres. A circle is the intersection of two spheres: if these invert into two new spheres, their intersection will invert into the intersection of the new spheres, which is a circle. The cases where one or more of the four spheres are planes can be considered similarly.

It will be noticed that RW after inversion becomes IOW. If we invert IOW, we come back to our original RW. We thus have a straightforward and easy way for passing from one to the other. This is all the geometry we shall need,

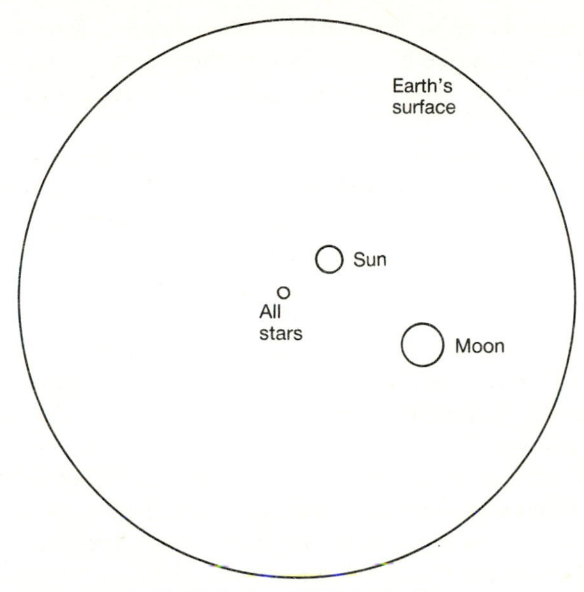

Fig. 4.4 The IOW world. Inside the hollow IOW Earth, the stars occupy a minute region 2.2 mm in diameter at its centre. The Sun has a diameter of about 2.5 m and lies at a distance of about 270 m from the centre. The Moon is about 960 m in diameter and moves in a path about 106 km from the centre of the IOW earth. The figure is grossly out of scale: if it were drawn correctly, Sun, Moon and stars would all be invisibly small.

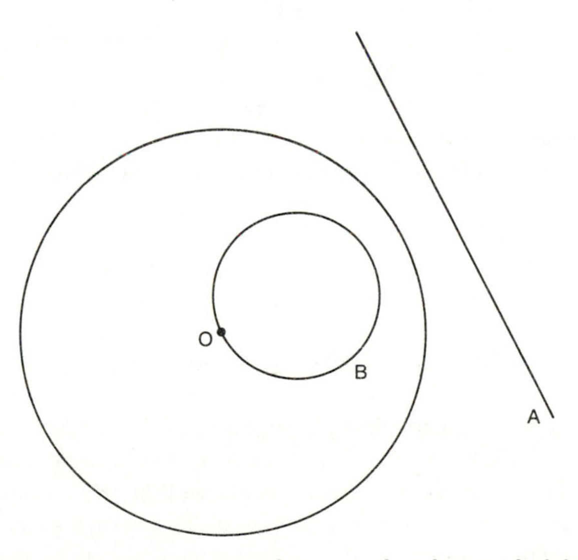

Fig. 4.5 The plane represented by A transforms into the sphere B which has the centre O on its surface.

and we complete the definition of IOW in the following way. Let any event occur in IOW. To describe it, we transform the situation from IOW to RW, and apply the ordinary physical description to the event in RW. Then we transform back to IOW.

For example, we throw a stone in IOW: how does it behave? We transform the original position of the stone in IOW to the position which corresponds in RW. We also transform the initial velocity of the stone in IOW to the corresponding velocity in RW. (The velocity can be represented by two points on the path of the stone in IOW separated by a short interval of time, say one thousandth of a second. These two points invert into two points in RW, which define the direction and speed of the stone in RW.) Then we apply Newton's laws to find the path of the stone in RW. Each point on this path inverts to give a point on the IOW path at the corresponding time.

This very brief sketch can easily be elaborated to cover other situations. It will be clear that we do not have to keep referring from IOW to RW. With a little effort, we can work out 'laws of nature' which apply directly to IOW. There will, for example, be something corresponding to Newton's first law, which will have the following form in IOW:

> Every body continues in a state of rest, or of motion along a circular path passing through the centre of the earth with speed proportional to the square of the distance from that centre, unless it is compelled by impressed forces to change that state.

There will also be a new law in IOW corresponding to the inverse square law for gravitational attraction in RW, and so on.

We now assert that the picture of the world represented by IOW is scientifically equivalent to RW: every observation which refutes one will refute the other, every observation which fails to refute one will fail to refute the other. That this must be true will be evident from the way that our description of events in IOW was defined, but the fact may nevertheless be surprising.

For example, it may be objected that we cannot be inside a hollow earth, because then ships would not sink below the horizon. But we have to remember that in IOW, light travels along circular paths passing through the centre of the earth. These have a greater curvature than the surface of the sea, and ships will disappear below the horizon in the usual way (Fig. 4.6).

Again, it may be objected that photographs of the earth have been taken from space, and these clearly show it as a convex spherical body. But the explanation follows in the same way as before (Fig. 4.7). Construct all the circular paths in IOW, representing rays of light which pass through the observation point A and the centre of the earth. Some of these will intersect the earth's surface, some will meet it tangentially, and some will fail to meet

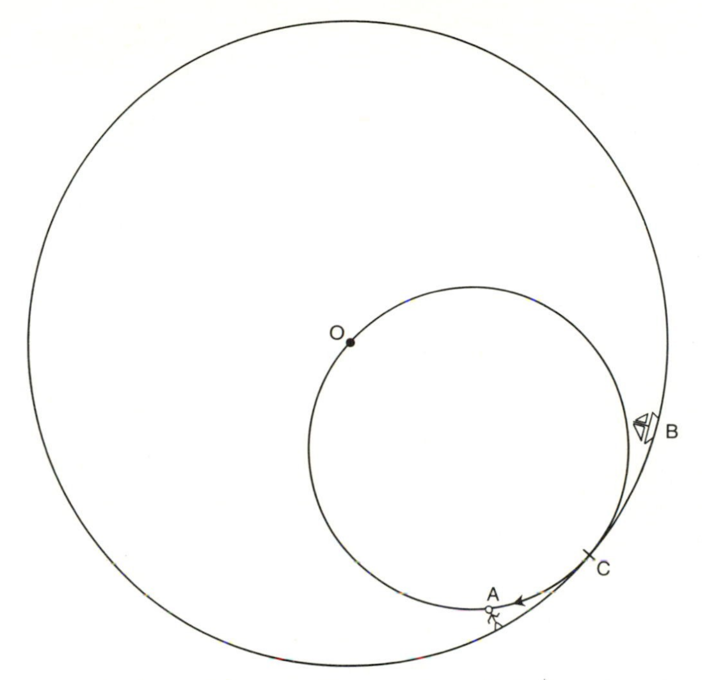

Fig. 4.6 With the observer's eye at A, there is a circle which passes through A and O and is tangential to the surface of the sea at C in IOW. Then C defines the horizon in a particular direction, and a ship at B will not be visible. The figure is grossly out of scale.

it. The effect, if we trace it in detail, will be exactly what is shown on the photographs.*

4. *Inferences and objections*

Accepting that IOW is indeed scientifically equivalent to RW in the sense we have defined, what should we conclude? There are a number of possibilities.

i. A very natural reaction would be that IOW is nonsense:

We know that the earth is not a hollow sphere. We know that it is a spherical body which moves around the sun on an elliptical path, accompanied by its moon. Certainly, with a perverse enough ingenuity we can construct chimerical theories to confuse the matter, but no reasonable person will accept them.

* Something similar to the IOW was suggested about fifty years ago, in a reference which I cannot now locate. We lived on the inside of a hollow sphere, but this was not equivalent to RW. For example it was suggested that measurements of the distance between two plumb-lines in adjacent mine shafts would show them to be further apart at the bottom than the top, which is the opposite of what we expect in RW. In IOW as defined here, the measurements would be exactly the same as in RW.

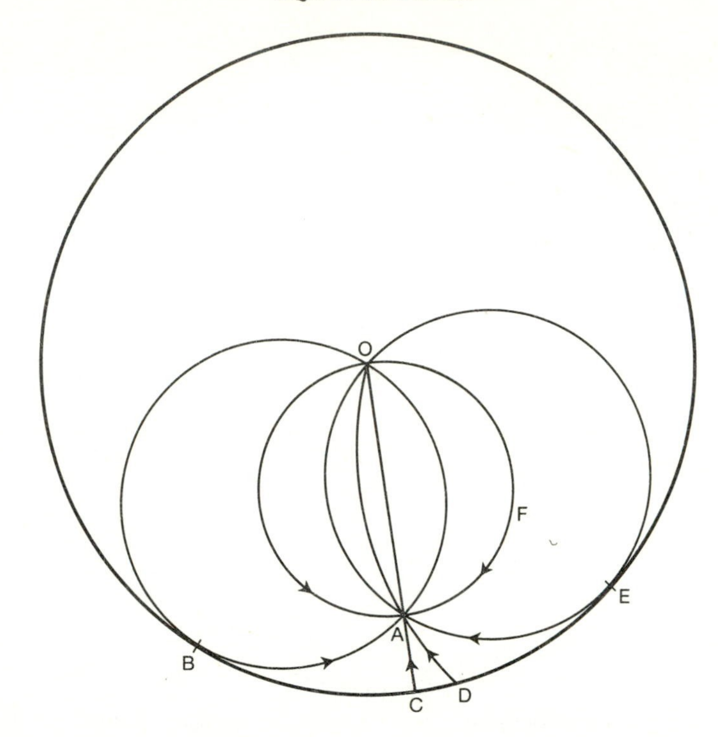

Fig. 4.7 Let the observer be at A in IOW, which is a point in space. The points C and D on the surface of the Earth will be visible by rays of light which are parts of circles passing through A and O. The points B and E are visible by rays which graze the surface of the Earth. Other rays such as F will arrive from space without being intercepted by the Earth, and by their agency the observer will see the stars and other astronomical bodies. The Earth will appear to the observer as a solid body floating in space.

To react in this way is rather useful. We find it difficult now to understand the heat and obstinacy with which the early seventeenth century rejected the Copernican view that the earth went round the sun, rather than the sun round the earth. Their reaction was very like the one just described, and to react in this way is to understand better what they felt. A very illuminating way of regarding Galileo's controversy with the Church is to say that he was putting forward a scientifically equivalent theory. The episode is not directly relevant to our present argument, which is logical rather than historical, so it is deferred until later (Section 8).

ii. If we can overcome the immediate rejection, attributing that to unfamiliarity, and consider the matter more coolly, we may still insist that IOW is nonsense:

Given the real world, we can certainly construct a form of words to describe an alternative, the IOW. But what we shall observe in the real world remains completely unchanged. Therefore the words which we use to describe IOW are without effect, and meaningless.

Unfortunately, this argument depends entirely on our prior acceptance of RW. If we already accepted IOW, the argument could be reversed to show that the form of words which we use to describe RW is empty and meaningless. The situation is entirely symmetrical between the two positions, so far as this argument is concerned.

iii. If we accept that RW and IOW are equally possible ways of describing the world, we may still wish to reject the second:

Granted that IOW can reproduce all the observed phenomena, it is unacceptable scientifically on at least two grounds: it is more complicated than RW, and it is anthropocentric. We are bound to choose the simpler explanation where we can. We are also bound to exclude ourselves from any privileged position in the universe.

These undoubtedly are strong motivations for the working scientist. To choose a more complicated explanation is aesthetically unpleasing, but also it is not a good way to make progress. The next stage in the development of a scientific explanation will be the more difficult to take, to the extent that the explanation is more complicated. As to anthropocentrism, that offends against historical attitudes deriving from the competition with religious explanations of the world. It also conflicts with the attempt to make our scientific explanations invariant under the widest possible change of circumstances. A race of beings inhabiting some other planet in the universe (assuming such to exist) would be unlikely to accept IOW, based upon our earth, as a way of describing their own experience.

These considerations undoubtedly operate, and they are sufficient to explain why in orthodox science we do not see equivalent theories coexisting on equal terms for long periods. One of the contenders will be judged superior on the grounds described, or on others of similar type, and will displace its rivals. Yet, as will be shown below, this procedure is an open door by which our preconceptions come to seem part of nature itself.

iv. We may consider that RW and IOW are not two theories, but one:

A theory is a description of the world which organizes our knowledge, allows us to predict the effect of actions, and to that extent permits us to control nature. In all these ways, RW and IOW give identical results. No observation can ever contradict one without also contradicting the other. So for all purposes of science (and, we may add, of technology) they are the same. The fact that we can imagine both RW and IOW, and can describe both of them mathematically, tells us something about the human mind, but nothing about the world beyond it.

This is the view we shall adopt, having first expressed it in different language. We shall say, using terms which have been explained earlier, that the appropriate thing to serve as a scientific theory is neither RW nor IOW, but the equivalence class to which they both belong. No two equivalence classes can themselves be equivalent (for then they would be the same class) so that the difficulty we have experienced in choosing between RW and IOW can never arise between two equivalence classes. There is always some observation which we can make which will produce different results in the two classes, and will therefore provide a scientific basis for discriminating between them.

5. *Theories and myths*

If we decide that the equivalence class including RW and IOW is the appropriate candidate to be called a theory, what are we to call RW and IOW themselves? It will be remembered that following Popper, we proposed that a theory should be treated as allowable in science if there exists some conceivable observation by which it can be refuted.

Both RW and IOW satisfy this criterion. RW, for example, consists of all the accepted scientific theories which we use to describe our world: Newton's laws and their more recent refinements in physics, chemical and biological theories, and so on. This theoretical body of knowledge is susceptible to refutation at a multitude of points. Except by inadvertence, it contains nothing which does not satisfy this criterion. As IOW gives rise to exactly the same result as RW in any observation, the same can be said of it.

Yet, as we have seen, RW and IOW are not suitable candidates to serve as theories. Each can be refuted, but the statement that 'the world is really like RW and not like IOW' cannot be refuted, nor can its opposite. Once we have admitted that it is possible to set up equivalent theoretical structures such as RW and IOW, the criterion which we propose for a 'scientific theory' needs to be refined. We need to say that, in addition to being capable of refutation, a scientific theory must also not be equivalent to another theory.

This last requirement is achieved by using equivalence classes as our theories. We then need a name for the members of an equivalence class (such as RW and IOW) which are not themselves theories. The term we propose is 'myth'. This has some unwanted associations, but so would any alternative suggestion unless it were entirely new and invented. We do not intend, by calling them myths, to say that RW or IOW are untrue: in ordinary language both of them are true descriptions of the world, so far as this is scientifically knowable. By 'myth' we mean rather a description of events which carries with it implicit values, and may serve a social purpose.

That RW, as the accepted myth, does carry values is clear. One powerful reason for preferring it to IOW is that it is simpler. If we uniformly select myths from their equivalence classes on the basis of simplicity, it will appear that nature itself prefers simplicity. If we always reject anthropocentric descriptions of nature, then the observed facts, through their theoretical interpretation, will appear to confirm that mankind has no special place in the universe. If we never permit an explanation in terms of purpose, nature will appear to be devoid of purpose, and it is this proposition which our discussion has been intended to introduce.

Now it will be clear that any attempt to develop science by using equivalence classes as our theories would give rise to insoluble problems. Our explanations are in terms of the myths, not the equivalence classes which are the theories. The equivalence classes are in themselves too abstract and tenuous to guide our thinking. If we regard them as a multitudinous collection of myths, we shall on the other hand become mired in a proliferation of equivalent myths which will prevent all progress. But this difficulty is easily overcome.

Following the mathematical procedure, we can nominate one myth to represent its equivalence class. Then we can test the theory, the equivalence class, by testing the representative myth. If this is refuted, then so is every other myth which is equivalent to it; that is, the equivalence class is refuted. And if the representative myth is not refuted, then the theory (the equivalence class) is not refuted.

If, as is natural, we choose RW as the representative myth, it seems that we have come full circle. We use it as though it were the theory, though we accept that it is not. But in practical terms our scientific procedure will remain the same as it has always been: the discussion of equivalence classes seems empty and redundant.

Yet this is not so. We have drawn a clear distinction between the scientific theory and the myth which represents it. The myth carries with it a range of values which governed its choice from among competing myths. If we believed that the myth was indeed the theory, then these values would seem to be embedded in nature. If we accept that the theory is something quite different from the representative myth, then we shall not attribute to nature those values which are incorporated in the representative myth, but are not incorporated in other myths belonging to the same theory. If we can reach and firmly hold to this position, we shall have achieved a quite different outlook on science and its relation to the world.

6. *Some observations*

An objection which may be made to the preceding discussion is on grounds of triviality. It will probably be accepted that the inside-out world is a feasible way of describing our observations, but as we are most unlikely ever to use it, what benefit does it bring?

The aim has been twofold. First, the dearth of examples of equivalent myths (as we now call them) could lead to the belief that they are rare and exceptional. The method we have used to generate IOW can clearly be modified in an unlimited number of ways to generate an unlimited number of equivalent myths. We can impose different transformations, rotations, reflections, and so on. These are all geometric.

We can equally make transformations of masses and forces and other dynamical variables. What prevents us from doing so is not infeasibility, but a general disapproval of this kind of activity. It is regarded as unserious, as contributing nothing to science, as frivolous and generally reprehensible. A representative myth, chosen for good reasons of simplicity and rejection of anthropocentric or teleogical notions, will not be questioned on grounds other than scientific validity.

We have in effect questioned such representative myths on the grounds of the values which they carry with them, and which through their agency appear to be supported by the facts themselves. To raise a question on these grounds cannot be forbidden. Yet unless a distinction is made between theory and representative myth, the question will be rejected, because it will appear to be an attack upon scientific knowledge itself.

We wish to question the values incorporated in a myth, without questioning the validity of the theory to which it belongs. As carriers of values, scientific myths fulfil the same function as other myths: as systems of belief which meet a social need. One cannot live in society without myths, but the attitude adopted here is the mildly sceptical one that Occam's razor should apply: myths to which we give assent should not be multiplied beyond necessity. And the role of myth should be distinguished from the role of representing a scientific theory.

We therefore propose that the 'mythical' function of the representative myth—its role as a carrier of values—should be rejected while its scientific function—its role of representing a theory—should be accepted. The only way in which this can easily be done is to illustrate by means of some unfamiliar equivalent myths the change in our outlook that their adoption would produce. We shall do that in particular for the causal and the purposive myths which can represent our theories about nature.

A second objection could be made on the grounds that science is not in practice carried out in any such rigidly logical way as Popper's account, or alternative accounts, might suggest. Consequently the logical structure which we have suggested, of theories consisting of equivalent myths and subject to refutation, cannot represent the true nature of scientific knowledge.

This is readily admitted, but it does not damage the argument. We have already said that science, as a successful and self-confident activity, cannot easily be questioned from outside. What we have done, therefore, is to adopt an account of the scientific procedure, in an idealized form which will probably be acceptable to most scientists. They will probably admit that science as they know it does not conform to the ideal, but will probably agree that something which did conform to the ideal would be science.

Then we have shown that when this ideal picture of science is accepted there can be equivalent descriptions of nature (equivalent myths); that the equivalence classes are the true candidates to act as theories; that one myth may represent the theory; but that the values according to which this myth was chosen cannot be attributed to the theory. Yet that if we confuse the representative myth with the theory we shall give a description of nature in which the values of the myth appear to arise inevitably out of the facts.

This has been established by means of an example which we can readily admit to be trivial, but which has the merit of simplicity and obviousness. We can therefore enter on the more serious matter of purpose and causality with the framework of our discussion already established. The discussion of purpose will greatly strengthen the case we have made, but the case is established independently, and does not rest on the use we shall make of it in discussing purpose.

7. *The causal myth*

In the terms we have now established, the substance of Chapter 3 can be stated in the following way. Classical mechanics gives rise to two equivalent myths. One is the causal myth, represented by Newton's equations. The other is the myth of purpose, represented by Hamilton's principle. These, together with any other equivalent myths that there may be, form the equivalence class which is the theory of classical mechanics.

The orthodox view does not see things in this way. The causal myth is regarded as the theory, and the other is dismissed as interesting, and sometimes useful, but not fundamentally significant. Here, for example, is the verdict of one of the authoritative accounts of variational principles:

> To declare in earnest that there is some deeper meaning attached to the idea of nature's acting in such a way that it makes some quantity a minimum, is perhaps a relic of the times when the universe was believed to be driven by a superhuman being. Obsessed with the urge to achieve 'cognition of real nature' i.e., to gain an 'explanation' of material phenomena, physicists have assigned to a particle the knowledge which enables it to take the most convenient path of all those which would direct it to its destination. The Aristotelian final cause is dragged into the subject. . . .[5]

Apart from the confusion of schedule and policy which was noted in Chapter 3, this is a legitimate condemnation of those who would attribute the values incorporated in the purposive myth to the theory, and thence to nature. But there is no corresponding reticence with respect to the values of the causal myth:

> The belief in a purposive power functioning throughout the universe, antiquated and naive as this faith may appear, is the inevitable consequence of the opinion that minimum principles with their distinctive properties are signposts towards a deeper understanding of nature and not simply alternative formulations of differential equations [i.e., causal relations] in mechanics.[5]

One can profitably turn this statement on its head:

> The belief that the world and all that it contains is a machine without purpose, mechanistic and anti-human as this faith may appear, is the inevitable consequence of the opinion that causal relations with their distinctive properties are signposts towards a deeper understanding of nature and not simply alternative formulations of variational principles.

Nothing like the latter appears in the orthodox accounts of physics. The consequence is that the causal myth is taken for the theory. Nature is explained as a machine without purpose. Then the facts themselves seem to impose upon us inescapably a natural world in which purpose cannot exist: in which any appearance of purpose is an illusion arising from strictly causal behaviour.

The alienation and distress to which this view can give rise have been eloquently expressed by Monod.[6] He describes the orthodox view of evolution and the origin of life. As the earth cooled, chance encounters between atoms caused them to agglomerate occasionally and temporarily into molecules. Eventually a molecule was formed which was able to replicate itself; to cause its environment to make further copies of itself. Multiplying in this way, the molecule became abundant.

From time to time mistakes occurred in the process of replication, producing a different result. Most of these anomalous outcomes were abortive, but in the end, after an immense time, an error in copying produced a new molecule which also could replicate itself, and more successfully than the first. So, with due elaboration of the proposed explanation, there began the process of evolution. Species reproduced themselves continuously and accurately except for chance errors. Some errors produced new species which competed for resources with those already existing. Millions of years of this process have produced the world we know, with its plants and animals and men and women and bacteria and viruses.

The explanation is a causal one, and Monod does not shrink from its implications:

> In both their macroscopic structure and their functions, living beings . . . are closely comparable to machines.

Living beings are chemical machines.

Here by 'machine', Monod implies an absence of purpose and a purely causal operation. This consequence is seen as an inescapable result of our scientific attempt to understand the world:

The cornerstone of the scientific method is the postulate that nature is objective. In other words, the systematic denial that 'true' knowlege can be reached by interpreting phenomena in terms of final causes—that is to say of 'purpose'.

But to make dialectical contradiction the 'fundamental law' of all movement, all evolution, is still an attempt to systematize a subjective interpretation of nature, showing it to have an ascending, constructive, creative intent, a purpose; in short, to make nature decipherable and morally meaningful. . . . This interpretation is not only foreign to science but incompatible with it.

. . . all [ideologies] take an initial teleonomic principle as the 'primum movens' of evolution. . . . In the eyes of modern scientific theory all these concepts are erroneous . . . for factual reasons.

Pure chance . . . at the very root of the stupendous edifice of evolution: this central concept . . . is today the *sole* conceivable hypothesis . . . And nothing warrants the supposition that conceptions about this . . . ever could be revised.

There is no scientific position, in any of the sciences, more destructive of anthropocentrism than this one.

But a world conceived in these terms is unlovable and unlovely, and Monod feels this deeply:

. . . the choice of scientific *practice* . . . has launched the evolution of culture on a one-way path; on to a track which nineteenth-century scientism saw leading infallibly to a vast blossoming for mankind, whereas what we see before us today is an abyss of darkness.

Modern man is engaged in

an anxious quest in a world of icy solitude . . . like a gypsy, he lives on the boundary of an alien world; a world that is deaf to his music, and as indifferent to his hopes as it is to his suffering or his crime

Now first this is not 'scientism': it is no vulgarization of the scientific outlook but the deeply felt conviction of an eminent scientist. But secondly, it agrees exceptionally well with what was said earlier. Purpose and anthropocentrism are rejected a priori by science. Then the facts of nature, as interpreted by the scientist, confirm that nature itself rejects anthropocentrism and purpose. The sceptic may wonder whether anything is so true of nature as Monod suggests, except those things upon which we have determined before we begin to study and interpret it.

Let us suppose that what Monod describes is one myth, the causal myth, explaining evolution. Let us also suppose that there is a second myth, the purposive myth, which is equivalent to the first, and that the causal myth describes the policy which can be derived from the purpose. Then Monod is asserting that the causal myth is the theory, and all facts which support the theory then support the myth. The values which operated in the choice of the myth reappear as inescapable consequences of the facts.

On this view, the alien, mechanistic, inhuman characteristics of the world which Monod confronts were brought to it by Monod, and by his fellow scientists in preceding centuries. They determined that science should reject purpose, should be 'objective' to use Monod's word. They also determined that science should reject anthropocentrism. Where a choice of equivalent theories presented itself, they chose according to these criteria, and anthropocentrism and purpose disappeared from science and from the world. Monod's tragedy is turned to comedy, and there is a wry, faint echo of Molière, 'Vous l'avez voulu, George Dandin.'

This is not to suggest that any conscious choice was made between equivalent myths: conscious choice would have broken the apparent inevitability of the conclusion. Nor would the mathematical apparatus for studying purpose have been fully available in earlier times: it has arisen in the past decades in response to technological needs, though Hamilton and his predecessors show how it could have arisen within science.

What seems to have happened is that the conflict between Galileo and the Church led to an uneasy division of knowledge. Purpose was claimed by the Church, and causality was left as the domain of science. Subsequently the preconceptions of causality embodied in science inhibited any large and serious effort to explore the possibility of an alternative. Fashion, and the consensus of scientific opinion, have made one line of thought acceptable, and other lines unacceptable. Hamilton's work was treated as an interesting mathematical exercise, but explicit warnings were given against regarding it as a contribution to our understanding of the world.

8. *Galileo: a historical example*

The clash between Galileo and the Church in the early seventeenth century is commonly seen as marking a break with the theologically determined outlook of the middle ages, and the beginning of the empirical, scientifically orientated view of the present time. The episode has traditionally been presented in a schematic form, which showed Galileo as the defender of free enquiry against the obscurantist dogmatism of the Church:

> A man is here revealed who possesses the passionate will, the intelligence, and the courage to stand up as the representative of rational thinking against the host of those who, relying on the ignorance of the people and the indolence of teachers in

priest's and scholar's garb, maintain and defend their position of authority. His unusual literary gift enables him to address the educated men of his age in such clear and impressive language as to overcome the anthropocentric and mythical thinking of his contemporaries and to lead them back to an objective and causal attitude towards the cosmos, an attitude which had become lost to humanity with the decline of Greek culture.[7]

This traditional account has been modified by more recent study[8]—and the picturesque fiction of Galileo's torture and imprisonment has been removed—but its influence lingers in the popular view that Galileo was 'right' and the Church was 'wrong'. If the preceding account of theories is accepted, showing them as equivalence classes of myths, a different interpretation of the episode can be given. Like the traditional view it is schematic, because a historical event is always richer than any reasoned

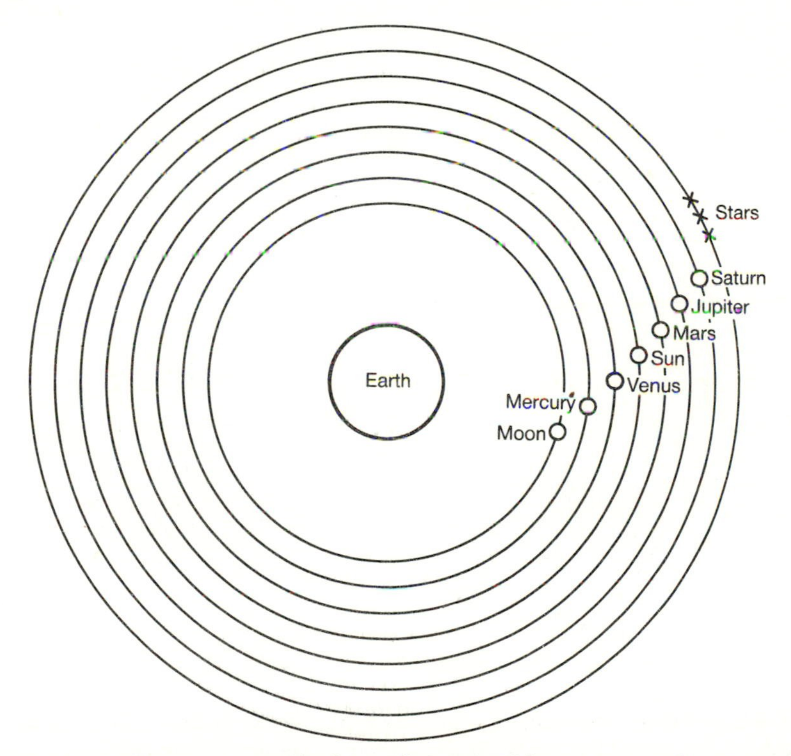

Fig. 4.8 The Ptolemaic universe. Earth is at the centre of the universe, and is surrounded by an outer sphere carrying the stars, rotating with a daily motion. Seven other spheres share in the motion of the outer one, but also move slowly in relation to it, in order to produce the apparent motion of the Sun, Moon and planets with respect to the stars. Reconciliation of the observed positions of these bodies required a complicated machinery of circular motions superimposed upon eccentric circular motions, but these were of interest only to astronomers.

account of it which we can give. Nevertheless, it contains some essential features which illustrate the role of myths as carriers of values.

In the early seventeenth century, three different accounts of the universe were current: the Ptolemaic, the Copernican, and the Tychonic. In the Ptolemaic system, which had long been the favoured one, the Earth was at rest within successive moving spheres, Fig. 4.8. These carried with them in their motion the moon, Mercury, Venus, the Sun, Mars, Jupiter, Saturn and the stars, giving them their daily and annual movements. To reconcile the positions of these bodies with observation (that is, to 'save the appearances'[9]) a complicated series of eccentrics and epicycles was introduced, but it was not necessary to believe in the physical reality of this cumbersome machinery. In Galileo's words, it was 'merely assumed by mathematical astronomers in order to facilitate their calculations.'[10]

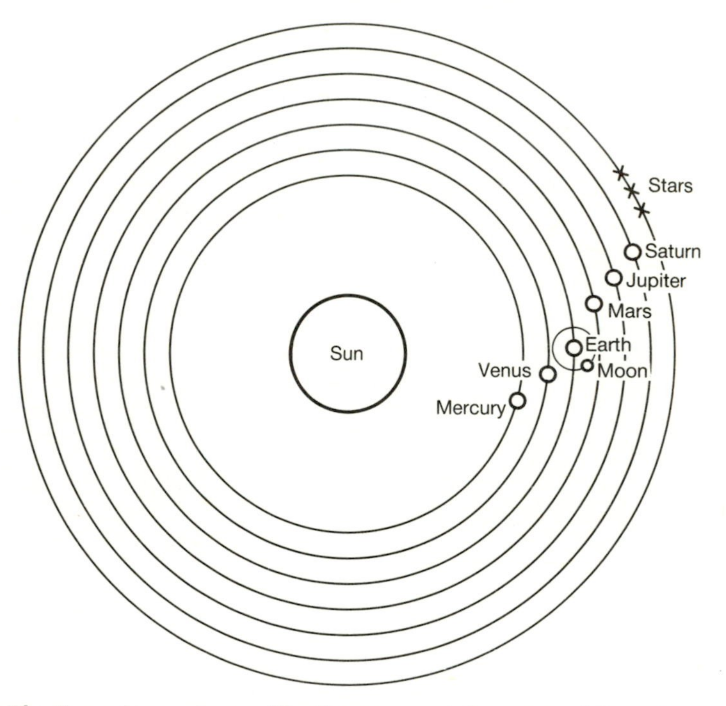

Fig. 4.9 The Copernican universe. The Sun is now at the centre of the universe. The Earth rotates around the Sun in an annual course, but also rotates daily upon its own axis, giving the apparent daily motions of Sun, Moon, planets and stars. Because Venus remains close to the Sun in the sky, it can never show a fully illuminated disc in the Ptolemaic universe; but it can do so in the Copernican system because, like the other planets, it has a slow movement around the Sun. Reconciliation of the observed positions again involved the complication of epicycles and eccentrics, which were finally eliminated by Kepler's discovery of elliptical motion, later derived by Newton from the inverse square law of gravity.

In the Copernican system, Fig. 4.9, the Sun was at the centre of the universe. Mercury, Venus, Earth, Mars, Jupiter and Saturn revolved around the Sun, while the Moon revolved around the Earth, the Earth rotated on its axis, and the stars were fixed. Eccentrics and epicycles were again used, but some conceptual simplification was nevertheless achieved. In principle, the Copernican system should have been distinguishable from the Ptolemaic: in the former, but not in the latter, Venus should have exhibited phases like the Moon's, while the stars should have moved slightly with respect to one another as the Earth went round the Sun. Neither of these effects could be observed: this was set down correctly to observational difficulties, though the explanation tended to weaken the credibility of the Copernican system.

The third system was the most recent, and was due to Tycho Brahe. We can imagine it as being obtained by superimposing on the whole Copernican universe a motion equal and opposite to that of the Earth.* The Earth was thus brought to rest at the centre of the universe, the Sun and Moon revolved around the Earth, and the planets revolved around the Sun. Conceived in this way, the Tychonic system was simply the Copernican system as seen by an observer on Earth who considered himself to be at rest: it was geometrically indistinguishable from the Copernican system by any observation. No celestial dynamics had yet been worked out, but if it had existed for one system, it could readily have been transferred to the other by the device used in Section 3. We can therefore regard the Copernican and the Tychonic systems as equivalent myths.

The situation, therefore, was that there were two theories. One was represented by the single, Ptolemaic, myth. The other contained two equivalent myths, the Copernican and the Tychonic. In principle, evidence which would decide between the theories was known—for example the existence or otherwise of phases of Venus—but it was beyond the power of existing methods to observe it.

In the period before the systematic testing of hypotheses became well established, the simultaneous existence of competing descriptions of the world was more common than it has since become. The situation was therefore familiar to the Church, which had a well-developed attitude to it. The Bible was true by inspiration, but its purpose was not to teach physics or astronomy. As St Augustine expressed it,

> One does not read in the Gospel that the Lord said: I will send to you the Paraclete who will teach you about the course of the sun and moon. For he willed to make them Christians, not mathematicians.[11]

* There were, in fact, some discrepancies from an exact interpretation of this scheme. For example, the stars were not given the motion which would have corresponded to the earth's movement around the sun, because relative motion between earth and stars could not be observed then, or for long afterwards. We ignore these historical differences, which were easily removable and logically inessential.

Because

two truths cannot contradict one another, it is the function of wise expositors to seek out the true sense of scriptural texts.[12]

That is, if the accepted interpretation of Scripture seems to contradict observed fact, the interpretation must be adjusted:

since Holy Scripture can be explained in a multitude of senses, one should adhere to a particular explanation only in such measure as to be ready to abandon it if it be proved with certainty to be false.[13]

On the other hand, in the absence of positive evidence against it, the traditional interpretation of Scripture was to be accepted. Cardinal Bellarmine summed up the situation in this way:

I say that if there were a true demonstration that the sun was in the centre of the universe and the earth in the third sphere, and the sun did not go around the earth but the earth went around the sun, then it would be necessary to use careful consideration in explaining the Scriptures that seemed contrary. But I do not think there is any such demonstration, since none has been shown to me . . . and in case of doubt one may not abandon the Holy Scriptures as expounded by the Holy Fathers.[14]

In the terms which have been introduced earlier, we may describe this position in the following way. Between equivalent myths, belonging to the same theory, the Church claimed the right to nominate the representative myth in accordance with the Scriptures. Between different theories, that is, different equivalence classes, the Church did not profess to judge. The matter was to be settled by the scientific evidence, and if the result appeared to conflict with the Scriptures, then these were to be suitably interpreted. It might, for example, be said that the Bible used familiar language about the heavens to teach the truths with which it was concerned, without intending a literal definition of the nature of the universe.

Between 1610 and 1612, Galileo, using the newly invented telescope, observed the phases of Venus, and the Ptolemaic system was thereby made untenable in its existing form for those who accepted this evidence. One theory, represented by the Ptolemaic system, was falsified.* The other contained the Copernican and the Tychonic systems, which remained equivalent and survived. Galileo ignored the Tychonic system, as he did consistently, and set up an opposition: if the Ptolemaic system is false, then the Copernican system is true. Dreyer comments on Galileo's publication in 1632 of his *Dialogue*,

In the whole book there is no allusion whatever to the Tychonic system, although it

* Kuhn's claim[15] that falsification is never accepted at once as clear and absolute refutation is surely justified, and is illustrated by the historical account, but in hindsight we may see the matter as it is described here.

is scarcely too much to say that about the year 1630 nobody, whose opinion was worth caring about, preferred the Ptolemaic to the Tychonic system.[16]

The dispute, in these terms, was therefore about the right to nominate the representative myth. The Church preferred the Tychonic system, in which the earth was at rest. Galileo insisted on the Copernican, in which the Earth moved. From the point of view given earlier, the mobility or immobility of the earth was not a property of the equivalence class, which alone could be empirically tested, but of the two equivalent myths which it contained and which no conceivable experiment could separate empirically.

The criteria of choice which the two sides wished to use were totally opposed to each other. To the Church, the world was a stage provided for the Christian drama of the Creation and Fall and Redemption of mankind. It would therefore choose, from among equivalent myths, the account which provided the fittest setting for this intensely poetic and dramatic story. Moving the earth from the centre of the universe irretrievably damaged the setting, as we see if we attempt to transpose Milton's *Paradise Lost* from the Ptolemaic to a Copernican, or still more a Newtonian, universe.

Galileo's criteria, on the other hand, are close to those of modern science. 'Nature,' he holds, 'takes no delight in poetry.'[17] On the contrary,

> The natural world was now portrayed as a vast, self-contained mathematical machine, consisting of motions of matter in space and time, and man with his purposes, feelings and secondary qualities was shoved apart as an unimportant spectator and semi-real effect of the great mathematical drama outside.[18]

To Galileo, the virtue of a hypothesis lay in its power of explanation, and the guidance which it gave towards further understanding.

Galileo was deeply engaged in the attempt, completed by Newton, to give a dynamical theory of the heavens. His judgement that a heliocentric universe was the appropriate one for developing this theory was entirely sound. The alternative development of dynamics which would have been required by the Tychonic myth would have been so much more complicated as to make its emergence unlikely.

But equally, we can see in retrospect that the Church was right, in believing that the choice of one myth to represent the theory, rather than the other, carried far-reaching consequences, going beyond its effectiveness for the development of science. As subsequent history has shown, confusion between the representative myth and the theory would ensue. Then the values according to which the myth was selected would come in time to seem properties of the natural world, objectively present and supported by all the facts we observe. We should have begun the process by which we have created for ourselves Monod's 'alien world' with its 'icy solitude' and its 'abyss of darkness'.

9. *Further course of the argument*

What has been said above defines the argument that we wish to put forward, but it goes a long way beyond what we have established. Hamilton's principle provides a purposive myth for a wide area of classical mechanics. We have pushed the arguments based upon it into the area of evolution, which rests upon chemistry, and chemistry is not underlain by classical mechanics but by quantum mechanics. We have also said nothing of the other main extension of classical mechanics, to relativity.

The next chapter is devoted to these topics, and we shall then return to our major preoccupation, which is the influence that has been exerted by the causal myth upon our understanding of the world and our behaviour in it.

References

1. Barry Barnes (1974). *Scientific knowledge and sociological theory*, p. 154, Routledge and Kegan Paul.
2. Karl Popper (1959). *The logic of scientific discovery*, Hutchinson; (1963). *Conjectures and refutations*, Routledge.
3. Thomas S. Kuhn (1962). *The structure of scientific revolutions*, Harvard University Press; Paul Feyerabend (1975), *Against method*, NLB; Imre Lakatos (1976), *Proofs and refutations*, Cambridge University Press.
4. For a reasoned account, see Mary Hesse (1980). *Revolutions and reconstructions in the philosophy of science*, Harvester Press.
5. Wolfgang Yourgrau and Stanley Mandelstam (1968). *Variational principles in dynamics and quantum theory* (3rd edition), pp. 173, 174, Pitman.
6. Jacques Monod (1970), *Chance and necessity* (translation 1971), pp. 30, 46, 47, 51, 82, 110, 158, 159, 160, Collins.
7. Albert Einstein (1953). Foreword, in Galileo Galilei (1932). *Dialogues concerning the two chief world systems—Ptolemaic and Copernican* (translated by Stillman Drake), University of California Press.
8. See J. L. E. Dreyer (1906). *History of the planetary systems from Thales to Kepler*, Cambridge University Press; Lane Cooper (1935). *Aristotle, Galileo, and the Tower of Pisa*, Cornell University Press; Stillman Drake (1957). *Discoveries and opinions of Galileo*, Doubleday; Thomas S. Kuhn (1957). *The Copernican revolution*, Harvard University Press; Arthur Koestler (1959). *The sleepwalkers*, Penguin; Jerome J. Langford (1966). *Galileo, science and the Church*, University of Michigan Press.
9. Reference 8, Langford, p. 29.
10. Reference 8, Stillman Drake, p. 26.
11. Reference 8, Langford, p. 65.

12. Galileo, in Reference 8, Stillman Drake, p. 186.
13. St Thomas Aquinas, in Reference 8, Langford, p. 66.
14. Reference 8, Stillman Drake, pp. 163–4.
15. Thomas S. Kuhn (1962). *The structure of scientific revolutions*, Harvard University Press.
16. Reference 8, Dreyer, p. 416.
17. Galileo, quoting Guiducci, in Reference 8, Stillman Drake, p. 238.
18. Burtt, quoted by Langford, Reference 8, p. 167.

5 Extensions of the argument

Relativity, quantum mechanics and subordinate purposes

1. *Special relativity*

The outline of our argument has now been sketched in. It is possible to have equivalent descriptions of the world (equivalent myths) which cannot be distinguished by empirical evidence. If we choose one of these, on some grounds of preference, and say that it is the one true and valid description of the world, then this description will be supported by the physical evidence. Our preferences will then seem to arise objectively out of our experience of the world. Further, we have contended that the preference for efficient causes rather than final causes in our explanations provides an example of this procedure.

The argument has been developed in the briefest possible way, so that the outlines should not be obscured by detail. In particular, the equivalence of purposive and causal explanations was based on Hamilton's principle. In the form in which it was stated, this applies only to classical systems, and the restriction, if it were not removed, would throw doubt on the general validity of the point that is being made. We shall therefore give extensions to quantum mechanics, and to relativity.

Of these extensions, the second is the easier because the work has already been done.[1,2] There are two versions of relativity theory—special and general—and we first consider special relativity. This grew from difficulties which were found in the classical theory, towards the end of the nineteenth century, in dealing with the velocity of light.

If we consider light to consist of particles, we should expect their apparent speed to depend upon whether we are travelling towards or away from their source. If on the other hand we consider it to consist of waves in some medium, we should expect to discover different speeds of light in different directions when we are moving with respect to the medium. Neither effect can be found: the speed of light always has the same value regardless of its source and regardless of our motion.

This difficulty was overcome by Einstein's special relativity in 1905. As interpreted by Minkowski, it requires us to give up our usual sharp distinction between time and space, and to regard them as to some degree interchangeable. We add time to the three dimensions of space, and regard the four variables as four dimensions of a new 'space-time'. Relative motion of two observers involves a rotation of one space-time with respect to the other, in which time can be partly converted to space, and space to time.

The details of this need not concern us. From our point of view, the important thing is that something entirely analogous to Hamilton's principle still holds. What happens is simply that the formula corresponding to 'rate of consumption of fuel' is changed. In classical mechanics this was the Lagrangian, which was given by the kinetic energy, less the potential energy. In special relativity, the Lagrangian has a different form.

In explaining this, it is convenient to measure physical distances in 'light-seconds', where a light-second is the distance travelled by light in one second. This is about 299 800 km, and with this unit of length the speed of light obviously becomes one unit per second. In relativity, energy and mass become interconvertible, so we add the potential energy to the mass of a particle. Then the 'rate of consumption of fuel' is represented by this sum of mass and potential energy, with reversed sign. The rate is now not 'per second', but 'per unit of distance' measured in the four-dimensional space-time, and the 'total fuel used' has to be stationary.

This sounds at first like something very different from Hamilton's principle. But in fact, for speeds which are small compared with the speed of light, the two give very nearly the same result. Only for particles moving at an appreciable fraction of the speed of light do the two principles begin to diverge.

We shall not attempt to construct an analogy, like our ship, for this new principle. It is enough for our purposes that it has a comparable form. Just as in classical mechanics, we can regard the variational principle as giving a description of relativistic systems in terms of purpose. There is, as before, a causal description in terms of differential equations which constitute the policy, and the two descriptions are equivalent.

2. *General relativity*

In special relativity, the effect of gravity is represented by a contribution to the potential energy. General relativity eliminates the need for this: gravity is taken into account by a modification of Minkowski's four-dimensional space. Whereas in special relativity this space is 'flat', in general relativity it becomes 'curved'.

The idea of a curved two-dimensional space is not difficult to appreciate.

The surface of a sphere is an example, and the curvature is evident because the sphere is imbedded in our ordinary three-dimensional space. But a space can be curved without being imbedded in a higher-dimensional space, curvature being an inherent property of the space itself.[3] Unfortunately this becomes impossible to visualize, though it can be developed mathematically in a consistent way.

In general relativity, a mass causes a curvature of four-dimensional space-time. In this curved space, another mass under the gravitational influence of the first follows a path between two points which is defined in a particularly simple way: it is one having stationary length, so that any small change in the path between the points leaves the length unchanged. The 'points' defining the ends of the path are in four dimensions, and so specify the position and time at the beginning and end of the path.

Again we have no need to go into details. It is sufficient to know that in general relativity there is a description in terms of purpose. There is, naturally, a causal description which can be obtained from the purposive description and is equivalent to it.

An interesting feature of general relativity is that the purposive description takes the central place. It gives a simplicity and elegance to the theory, in spite of the mathematical details, which are appealing. One would think that this example might have removed some of the philosophical objection to descriptions in terms of purpose, but that is not so.

3. *Systems with random disturbances*

The second major development in twentieth century physics, besides relativity, is quantum mechanics,[4] which arose from difficulties in explaining the behaviour of very small particles. One feature of these is an inescapable randomness in their behaviour. For a classical particle, we can in principle specify the initial position and velocity with arbitrarily high accuracy, and then its position and velocity at all future times can, again in principle, be calculated. We say 'in principle' for two reasons. First, our practical means of measurement have limited accuracy, so we cannot know the initial position and velocity with the precision that would be needed for exact calculations. But even if this practical limitation could be continually relaxed, there would come a point where we should be unable to treat the particle as classical, to the accuracy needed, and would have to treat it as quantum-mechanical.

For quantum-mechanical particles, even in principle, there is an absolute barrier to the accuracy with which we can know simultaneously the position and velocity. Then at future times, all we can do is to calculate, for example, what is the probability that the particle will be found at a particular position. It should be noted that though we have talked of 'classical particles' and 'quantum-mechanical particles', these are not different things. All particles, strictly speaking, must be treated by the methods of quantum mechanics. But often a classical description, which is simpler, is so good an approximation that we use it instead.

Now unfortunately, in quantum mechanics there is no variational principle which plays the same major role as those we have mentioned before: Hamilton's principle, with its extensions for special and general relativity. As Yourgrau and Mandelstam[5] say, 'Generally speaking ... stationary principles do not occupy in quantum mechanics the prominent place they hold in classical mechanics.' If this were an inherent and irremovable feature, it would destroy the generality of our main argument. We should be able to say that in classical mechanics and relativity there are two equivalent descriptions, based respectively on purpose and causality. But in the major and basic area of quantum mechanics only one, the causal description, would be available. The conclusion would have to be that descriptions in terms of purpose cannot be regarded as general, and equivalent, alternatives to causal descriptions.

Some will question whether quantum mechanics can legitimately be described as 'causal', and we shall return to this point in Section 8. Accepting for the moment that it can, the question that arises is whether an equivalent purposive development of the theory can be given. In view of the immense effort which has been applied to the development of the existing, orthodox theory, there might seem to be little hope of discovering an alternative which had previously escaped notice. But there are two reasons for thinking that this conclusion would be too pessimistic.

The first is that the mathematical development of variational principles in the presence of random disturbances is recent, and has taken place in control theory in the last forty years.[6] It has in recent years become known to some physicists, but not to many. The second reason is that the general, philosophical, objection to explanations in terms of purpose has probably limited further the effort devoted to a search for them.

There has nevertheless been a growing interest in recent years in the application of control theory to quantum mechanics, both among physicists and control theorists. The major line of development springs from work in physics by Nelson,[7] and it has been continued partly by the aid of control theory.[8] This approach suffers from a particular difficulty which makes it unsuitable for our present purpose: if it is regarded in the obvious way as the policy arising from a variational principle, then the principle is not non-anticipative. That is to say, it makes our present actions depend upon future information which we do not yet have. One cannot condemn a theory absolutely on these grounds if it is successful in other respects, but certainly it is an unwelcome feature. It makes impossible, of course, any analogy with familiar situations such as we have used before.

Accordingly a different and rather simple approach will be used, which has been described elsewhere[9] and is sketched in outline in Appendix 2. This also has its difficulties, as might be expected, but they are not of a kind prevent entirely the familiar analogies which we wish to make.

In preparation for the variational principle in quantum mechanics we first sketch some features of the control of systems in which random disturbances are encountered. The theory goes under the name of 'stochastic optimal control', but it would be too pretentious to use this title for the little we have to say, and a more modest section heading will be used.

4. *Optimal control with random disturbances*

Let us return to our analogy of a boat sailing upon the ocean, subject now to random disturbances. We shall shortly consider how to set up a variational problem, and how to obtain the policy from this. Assuming that we have the policy, let us first consider what the situation will be. The policy will give a causal description of the behaviour which is necessary to satisfy the variational principle. Given the position and the time, it will specify the heading and speed required.

Now in the absence of disturbances, we could (at least in principle) obtain our present position exactly by dead reckoning, using the starting point and our speed and heading up to the present time. When disturbances exist, as we now assume, this is no longer possible. We know our heading, and our speed through the water, but the water itself is swirling and eddying in a random manner. Sailing on different occasions with the same heading, and speed through the water, we shall arrive at different points. We can form an estimate of our position if we have some statistical information about the disturbances, but the estimate will be subject to error.

To avoid the complications to which this leads, let us suppose that we can measure our position by some independent means, for example from radio beacons. Now we know our position at each time, and by consulting the policy we can decide what speed and heading are appropriate. The situation is shown in Fig. 5.1, which can be compared with Fig. 2.11.

With the simplification arising from an accurate knowlege of position, how do we set about defining a variational principle? It will no longer make sense to ask for the minimum fuel consumption on a single voyage. Our policy will tell us, when we have formulated and solved the problem, what speed and heading to set at every point on the voyage. But the course we shall follow will be subject to the vagaries of the disturbances in the water, and will be different on every occasion, Fig. 5.2. We shall pass through

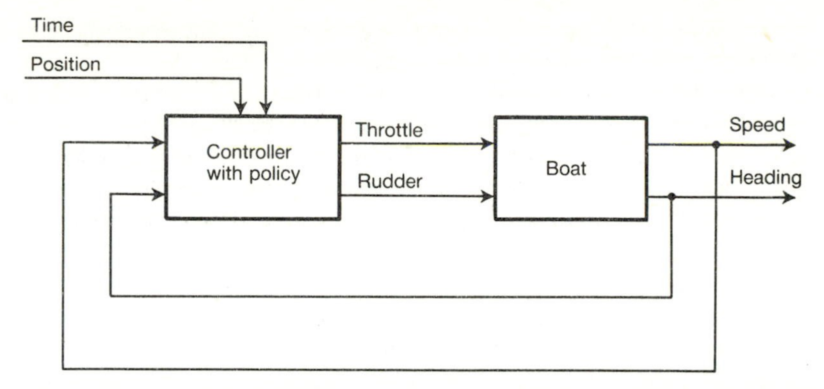

Fig. 5.1 Control system for automatic navigation when disturbances are present. When the sea is subject to random eddies, we can no longer obtain our position exactly by dead reckoning. We therefore suppose that independent means are provided for measuring our position.

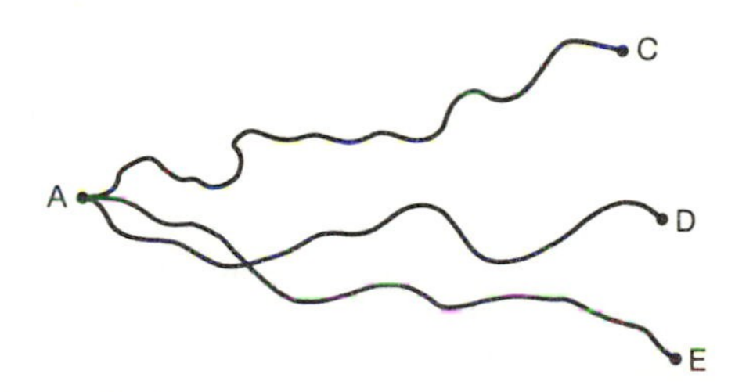

Fig. 5.2 We start from a fixed point A and follow a course determined by the control system in Fig. 5.1. Because of the random disturbances, we shall follow a different course on each occasion, and shall obtain different speeds and headings from the policy. After a fixed time, we shall reach different points such as C, D, E. These different courses will require different consumptions of fuel.

different regions, and use different values of speed and heading taken from the policy, and so shall use different amounts of fuel.

The obvious thing to do is to give up considering a single voyage, and to consider the average fuel consumption over a large number of voyages. We can suppose, for example, that there is a fleet owner who operates a large number of identical boats. He wants the total fuel bill for the fleet to be as small as possible. Given the particular properties of the disturbances, and the way the fuel consumption of the boats depends upon their speed and position, there will be some policy which defines the speed and heading to be used at each point and each time. If all boats follow this same policy, the fuel bill for the fleet will be as small as possible. How to find the policy is a question we will consider later.

Notice that in Chapter 2, where there were no disturbances, a boat following a policy traced out a unique, determinate path. We were therefore only interested in the policy along this path. Here the situation is different. Different boats following the same policy will pass through different regions, and we shall need the policy at all points which any boat may visit. With the kinds of disturbance which are usually assumed, there will in fact be some probability, though perhaps a very small one, that a boat will arrive at any accessible point whatever. We have to know the policy at every navigable point of the water.

One other question which has to be answered before we can look for a solution is, what constraints shall we place on the voyages? As in Chapter 2, some constraint is needed, otherwise a solution which gives zero fuel consumption is for the fleet to remain in port. (We are dropping the air conditioning requirement here and later.)

At this point we diverge from many technical applications of control, and from the discussion in Chapter 2. We began, in Chapter 2, by defining the starting point and finishing point of the voyage, and the time taken. By analogy, in the technical specification of an intercontinental ballistic missile subject to various disturbances, it might be required that the missile starting from a given point should land within a certain time inside a given circle with 95 per cent probability.

One way to deal with this kind of constraint is by a 'penalty function': to the cost of fuel used we add a fictitious cost, which is zero when the missile lands inside the target zone within the allowed time, but rises sharply with failure to achieve this aim. We find the policy which minimizes the new total cost, and by adjusting the severity of the penalty we can force 95 per cent of firings to meet the requirements.

We do not suppose that nature has any preferred targets calling for a penalty function, so we follow instead a procedure which was introduced later in Chapter 2. Instead of putting requirements on both ends of the trajectory, we impose all the conditions at the initial time. In Chapter 2, appropriate conditions at the initial time were the position and velocity (where as before velocity comprises speed and heading). What conditions are appropriate here will emerge later.

This completes the preliminaries, and we can now define our variational problem. We have a fleet of identical boats, and their fuel consumption depends on their displacement and on the square of their speed. They are to sail for a given time. They start with specified initial conditions, yet to be defined. Over the whole time, the fuel consumed by the fleet is to be a minimum. (More generally we should look for a stationary value as in Chapter 2, but it is easier to think and talk about a minimum.)

The development of the policy by Bellman's dynamic programming now proceeds in a way which is parallel to the method described in Chapter 2. In

Fig. 5.3, we consider for simplicity a one-dimensional problem: the boats are sailing to and fro along a canal. The horizontal axis shows time, from 0 to T, and the vertical axis shows position along the canal.

At a short time ε after the calculation begins, we assume that we know the optimal cost function (OCF). This is the minimum amount of fuel needed, from each position at time ε, for the remainder of the voyage: it is expressed as an amount per boat by averaging over the number of boats involved. We wish to find the best velocity to use at each position, in the interval of time from 0 to ε. With each point such as A we associate a velocity (with respect to the water) which will be maintained over the short time interval from 0 to ε. Because of the random motion of the water, boats starting from A and maintaining the same velocity will arrive at different points C, D, E, etc.

Knowing the statistical properties of the random disturbances, we can tell what proportion of the boats starting from A will arrive in a small region near a point such as C. There will be a certain consumption of fuel per boat (again averaged over the number of boats involved) to get from A to C, and a further consumption during the time from ε to T, given by the OCF at C. Adding these, and multiplying by the number of boats arriving in the region near C, we get a contribution to the fuel consumption during the time from 0 to T.

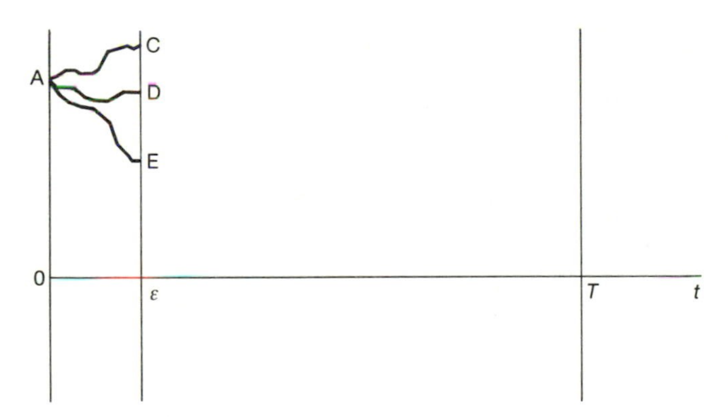

Fig. 5.3 One-dimensional problem in dynamic programming when there are random disturbances. A fleet of boats sails to and fro upon a canal for a time T, and is required to use the minimum amount of fuel. Distance along the canal is shown vertically, and time horizontally. Boats starting from A will arrive a short time ε later at various points such as C, D, E, and we can calculate what proportion of them go to each such point if we assume some policy giving velocity in terms of position at time zero.

If we know the optimal cost function (OCF) at each point at time ε, we can add up all the contributions to fuel consumption by boats starting from A. These contributions will consist of the fuel needed to get from A to points such as C, D, E, together with the minimum amount of fuel needed from C,D,E, etc., during the remaining time. Adding all contributions, we get the fuel used by the whole fleet in the time T.

Adding the contributions from D, E, etc., we get the consumption during the time from 0 to T for all the boats starting from A. This depends on the velocity we chose at A: by comparing the results of different choices we find the best velocity at A. As in Chapter 2, it seems that we need to know the answer, the OCF, before we can calculate it, but again this is not so. We can obtain an equation, admittedly a rather complicated one, which the OCF has to satisfy, and solving this equation gives us the OCF and hence the policy.

Before we can make the calculation, however, we must specify a constraint, as we did when there were no disturbances. There we began by saying what was the position at the initial time, and what was the position at the final time. Later, we replaced this constraint by specifying both position and velocity at the initial time.

When there are disturbances, as in the situation we are describing, we cannot reasonably ask for an exact position at the final time: the disturbances will prevent us from achieving it. In the general situation, an appropriate constraint is to specify the OCF (and hence the velocity) at the initial time, together with the distribution of the boats at that time. Then we can calculate the OCF at all later times, and from it obtain the optimal velocity, which generally will change with time. From this optimal velocity, and the initial distribution of the boats, we can calculate their distribution at later times, which again in general will change with time. When we speak of the 'distribution', we mean the number of boats in each small region, assuming that the number of boats in each region is large enough to make statistical fluctuations negligible.

The situation is simpler when the optimal velocity and the distribution of the boats remain the same at all times. We can, for example, imagine boats to be continually sailing from their harbours, and completing their journey at some other shore. At any given point, any boat will use the same optimal velocity regardless of the time when it arrives there. The number of boats in any region will remain unchanged (within statistical fluctuations) because as many will leave the region as enter it.

This condition represents a further constraint on the solution of the problem. When it applies, we can no longer choose the initial velocity and the initial distribution of the boats arbitrarily and independently: we can choose only one of these, and the other can then be calculated, though perhaps not uniquely. If, for example, we specify the velocity at every point at the initial time, then this velocity is maintained (at the same points) at all later times. There will then be only certain special distributions of the boats which will remain constant at all times when they use this optimal velocity. Similarly, a given distribution of the boats will remain unchanged with time only if the optimal velocity has certain quite special properties.

When we considered the motion of classical particles, our analogy with the motion of a boat satisfying a variational principle—minimizing the fuel—was an accurate one, subject to our assumptions. The motions of the boat and the classical particle could be made to agree exactly in numerical terms, by use of appropriate units of measurement. The motion of a fleet of boats subject to random disturbances, which we have given here, will serve in a similar way to illuminate some aspects of quantum mechanics. But it is no longer capable of being made exact, simply because quantum-mechanical particles do not behave in the same way as the classical objects to which we are accustomed. Nevertheless, some quite odd aspects of the behaviour of quantum-mechanical particles can be reproduced.

5. *Policies and schedules resumed*

The previous section is the deepest we shall need to sink into detail. With a few modifications it will allow us to see how a variational basis can be given for quantum mechanics. Before doing so, we look again at the difference between a schedule and a policy.

In the previous section, we assumed that time and position were always known. The best velocity (speed and heading) could then be obtained at each point and each time. This gave the policy, the causal relationship which implements the variational principle: if the position is such and the time is such (the causes) then the velocity must be such (the effect). It is a feedback, closed-loop solution because the velocity determines later positions (subject to the disturbance) and the position determines the velocity, Fig. 5.4.

If we had not supposed that the position was measured independently, we should have been unable to obtain a policy, and should have had to be content with a schedule. This gives the velocity at each time, regardless of position, and can be calculated before the voyage of the fleet begins. The calculation will be complicated, and we should probably try to split it up into two parts: make the best estimate of where we are, then use this in the best way to choose the velocity. The difficulty is that the two occurrences of 'best' are not independent: the 'best' estimate of where we are is the one which is best for the subsequent calculation of velocity.

In whatever way we obtain the schedule, it is clear that it will be worse than the policy, because it simply tells us: if the time is such, and the boat started from such a point with such a velocity, then the present speed and heading should be such and such. The schedule is calculated so that it gives the least fuel consumption for the fleet with the limited information available. With further information, namely the position of each boat at each time, we can do better because we know what was the effect of our past actions. This gives us the policy.

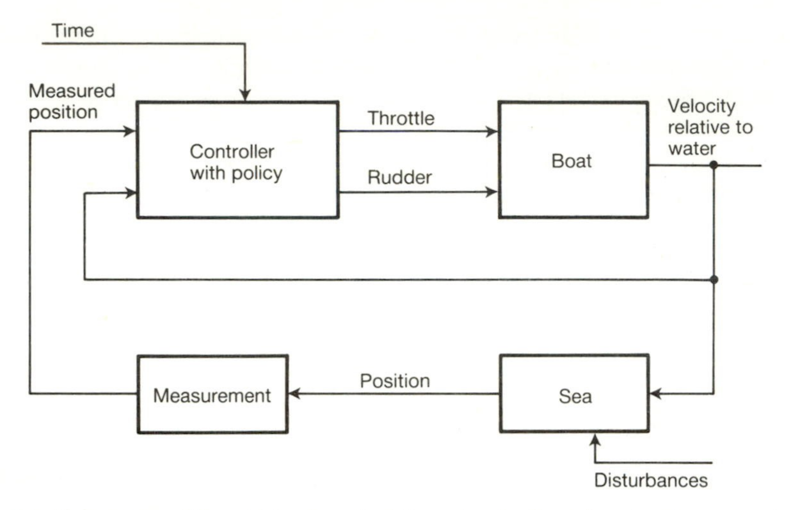

Fig. 5.4 Elaboration of Fig. 5.1 to show that there is a feedback loop involving position. The policy implemented by the controller produces a definite velocity with respect to the water at each time and each position. This velocity, together with the random disturbances, give the actual position, which is measured and fed to the controller. The aim of the policy is not to achieve a certain position, but to ensure the least consumption of fuel, for a fleet of identical ships, over the time from 0 to T. There is an inner control loop to ensure that the manipulations of throttle and rudder give the correct velocity (that is, heading, and speed relative to the water).

The schedule can be calculated beforehand, but it involves no knowledge of the future. We simply make the best estimate we can of what will happen, and act upon this. Our estimate will rely on things we know exactly—for example the starting point and time—and statistical information about the disturbances. The supposition that variational principles involve knowledge of the future arises from considering systems without disturbances. For these, the schedule gives the same result as the policy, and both allow the future to be predicted (in principle) exactly. But under the same conditions a causal description not obtained from a policy also (in principle) allows the future to be predicted exactly. The 'in principle' covers the reservations mentioned in Section 3.

The policy and the schedule just described are two extremes—the first based on complete knowledge of position and the second on a complete lack of direct information. Many intermediate situations can be defined, and appropriate controllers can be designed for them. As an example, position can be measured at intervals, as in the traditional practice of celestial navigation. Or we may provide a continuous independent means of determining position, say by radio navigation, but recognize that it is subject to random errors about which we have only statistical information.

In both cases, the result will be a fuel consumption intermediate between those of the policy and the schedule described earlier.

On the other hand, we can also consider situations in which less is known than we assumed when defining the schedule. Suppose, for example, that we are unsure of our starting point, and can only give some probability that we started from each of various different points. This uncertainty will further reduce the effectiveness of the schedule and will lead, for the fleet as a whole, to an increased fuel consumption. The important point in all of this is that the appropriate behaviour to fulfil the purpose depends on the information which is available, and changes as the extent of this information changes.

6. *The fishing fleet and the sandbanks*

In preparation for the following section, let us use what has been said to consider the situation shown in Fig. 5.5. A fishing fleet is based on the port A, and off the coast is a chain of low-lying sandbanks. Passages B and C exist between the sandbanks, lying over against A. Often both B and C are open, but because the sands shift with the tides, one or other of them may at times be closed.

Boats belonging to the fleet set out from A at various times, and under automatic control follow a policy which minimizes the consumption of fuel. They do not aim to arrive at any particular area, but fish as they go. The

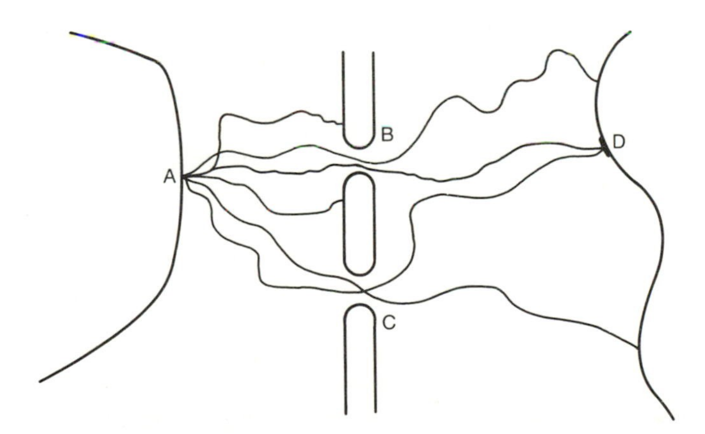

Fig. 5.5 Fishing fleet and sandbanks.
Boats belonging to a fishing fleet set out from A and sail according to a policy which minimizes fuel consumption. Some will pass through channel B, and some through C, if both are open. In a given period, a certain number of boats will arrive on a short section D of the opposite coast.

policy that they follow arises from an OCF which is calculated for the whole fleet over the complete sea area.

Some of the boats, sailing according to the policy, will arrive at a sandbank. They then abandon their voyage and fish where they are. Others will pass through the channel B, if it is open, or through C, and will continue their voyage beyond the sandbank.

When channel B alone is open, there will be certain values of the OCF beyond the sandbank, which we can calculate. We can draw the contour lines on which the OCF is constant. The boats will set a heading which is perpendicular to the contour line at the point where they find themselves, and a speed inversely proportional to the separation between local contour lines. This policy, together with the random motion of the sea, determines their progress.

When channel C alone is open, we can make a new calculation of the OCF, and it will be different from the previous one. Now we ask, how will the boats behave when both channels B and C are open? The answer will depend upon what information is available to the controllers on the boats, and we consider two cases.

i. If the on-board controllers have a record of which channel, B or C, they passed through, they can use the policy calculated when that channel alone was open. Those which pass through B will use the policy derived from the OCF calculated with B alone open. Those which pass through C will use the policy derived from the OCF calculated with C alone open.

Hence the behaviour of a boat passing through B will be the same whether or not C is open—the same, that is, so far as the policy is concerned, and so far as the statistical effects of the random disturbances are concerned. (Each separate voyage will of course have a different sample of the random disturbances, as in Fig. 5.5). In the same way, the behaviour of boats passing through C will be the same whether B is open or not. Notice that two different sets of OCF contours exist, and two different policies, and the boats use the one which is appropriate to them.

Now suppose that we count the number of boats which arrive in a certain time, say 24 hours, on a small stretch of coast D on a facing shore. We do this when B alone is open, when C alone is open, and when B and C are both open. Then we shall find that the third number is the sum of the first two, within some tolerance due to the random disturbances. The conclusion follows because we can conceptually separate the two sets of boats, one having used B and one C, even when they occupy the sea together. (We rule out collisions, of course.) These two sets have the properties which they would have if the other set were absent.

ii. Alternatively, suppose that the controllers on board the boats keep no record of which channel they passed through. All boats beyond the sandbanks will then have to use one single policy, derived from a single OCF. When B alone is open, this will be the one we calculated before: call it OCF(B). Similarly, when C alone is open we use the previous OCF(C). When both B and C are open, we shall have to compute a new OCF, say OCF(B and C), which will be different from OCF(B) and OCF(C). The policy derived from OCF(B and C) will then be different from the one derived from OCF(B) and from the one derived from OCF(C).

That is to say, when both B and C are open, boats in the region beyond the sandbank will steer a course which is different from the one they would steer at the same position and time if B alone were open, or if C alone were open. If now we count the number of boats arriving on the section of coast at D in 24 hours, we shall obtain a different result from the one we obtained in (i). In particular, this number will no longer be the sum of the numbers corresponding to B alone open and to C alone open.

7. *Quantum mechanics*

What has been said leads now in a natural way to the variational principle for quantum mechanics, with only one complication. As in Hamilton's principle, the 'rate at which fuel is used' is replaced by the Lagrangian, but we now add random disturbances which we did not have before.

It is the way in which we add the disturbances which provides the minor complication. It is easiest to explain if we consider 'simple harmonic motion', or SHM, which happens to play a specially important role in quantum theory. There are many common examples of SHM: the balance wheel of a watch, the pendulum of a clock, a weight on a spring, or alternating electric current. All of these oscillate with simple harmonic motion, Fig. 5.6.

Now mathematically, SHM is awkward to deal with because two velocities correspond to one position, as in Fig. 5.6. We can eliminate this difficulty by conceptually splitting the line along which oscillation occurs, so that motion to the right occurs in one part of the split line, and motion to the left on the other. The splitting could be done in any way, but the most convenient is to open the line out into a circle, Fig. 5.7. Then if we let a point travel round the circle with constant speed, the point vertically below it (or above it) on the diameter performs an SHM. At any point on the circle there is now only one velocity, and most of the properties of the SHM can be obtained conveniently from the motion of the point round the circle.

The horizontal axis in Fig. 5.7 is said to represent 'real numbers', while the vertical axis is said to represent 'imaginary numbers'. As we have introduced the latter from nowhere, for our own purposes, the name is appropriate, though it arose originally in a different way. Numbers which have a real part and an imaginary part are said to be 'complex'.

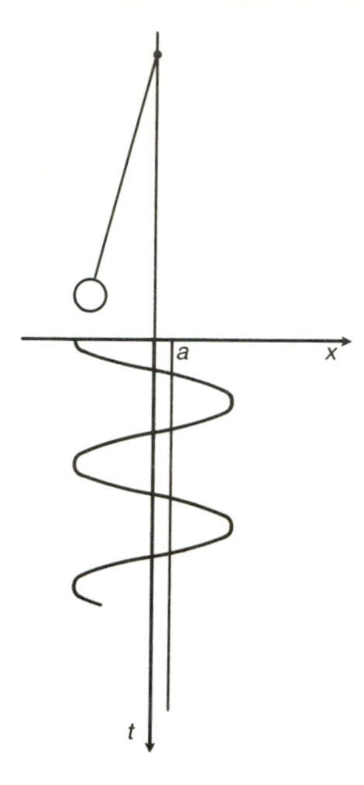

Fig. 5.6 Simple harmonic motion (SHM).
A pendulum (provided that it swings through only a small arc) performs, very nearly, a simple harmonic motion. Its displacement from the vertical position changes with time as shown. (The curve is sinusoidal.) At any position such as *a* during the swing, there can be two velocities: either towards or away from the vertical position.

Now it turns out that an appropriate way to modify Hamilton's principle, if we wish to get quantum mechanics, is to let every dimension become complex, and then to add random disturbances. As complex numbers behave in many ways like ordinary (real) numbers, this does not create much difficulty. The motion of particles in real space echoes the motion of points in complex space, as with SHM, though the individual points no longer correspond to individual particles.

We call the points in complex space the 'complex images', and can obtain their motion from an appropriate OCF, or more usually from the 'wave function', which takes its place in quantum theory and is simply a disguised form of the OCF. We may think of the complex images in most respects as analogous to the boats we have been considering. We have only to use some extra care, when we interpret the motion of the complex images to obtain properties of the motion of the particles in which we are interested.

Actually, since we have added random disturbances, we have to consider a 'fleet' of particles—this is usually called an ensemble. Corresponding to an ensemble of particles is an ensemble of complex images, and it is the behaviour of this which we compare to the behaviour of our fleet of boats. A

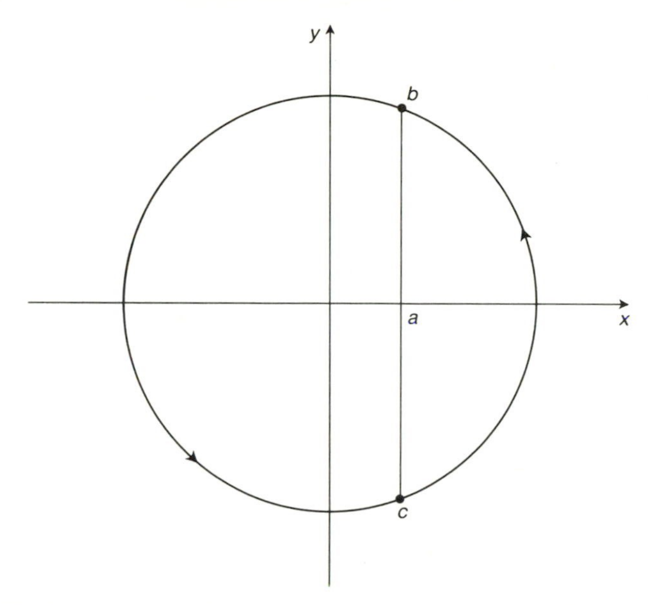

Fig. 5.7 The horizontal axis of x in Fig. 5.6 is opened up into a circle. If a point travels around this circle with constant speed, the point vertically below it (or above it) on the horizontal axis performs an SHM. The two different velocities at a correspond to two different points b, c on the circle. We represent distances on the horizontal axis by 'real numbers', and distances on the vertical axis by 'imaginary numbers'.

great many analogies could be drawn between them, but we shall consider only one, for which we have already prepared the way.

This concerns an experiment illustrating the phenomenon of interference, which is one of the cruxes in quantum theory.[10] Electrons are emitted from a source A, Fig. 5.8, and fall on a screen which has two holes B and C. (We could equally consider photons instead of electrons, but this would involve relativity, which we have not mentioned in connection with quantum theory.) One or other of B and C may be closed, or both may be open. Electrons which pass through the screen fall upon a second screen, and we count the number which arrive in a small region D in a certain time, say one second.

With B alone open, the number counted at D is, say, N(B), and with C alone open it is N(C). But with B and C both open, the number N(B and C) arriving at D is in general not the sum of N(B) and N(C). In fact we can find small regions D where N(B) is greater than zero and N(C) is greater than zero, but N(B and C) is zero.

We might at first suggest that electrons going through B have influenced those going through C in some physical way, for example by means of their electrical field. But we can reduce the flow of electrons until the probability

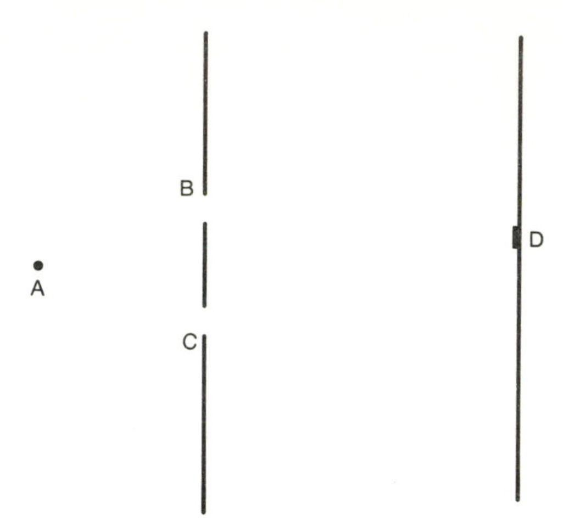

Fig. 5.8 A crucial experiment in quantum mechanics.
Electrons emitted from a source A can pass through hole A or hole B in a screen, or through both if both are open. With B alone open, we count the number $N(\text{B})$ of electrons arriving in, say, one second at a small region D of a second screen. In a similar way, when C alone is open we get $N(\text{C})$. When B and C are open, we get $N(\text{B and C})$, but this is not equal to the sum of $N(\text{B})$ and $N(\text{C})$. In fact we can find regions D on the second screen where $N(\text{B})$ and $N(\text{C})$ are both greater than zero, but $N(\text{B and C})$ is zero.

of having two electrons in the system at the same time is vanishingly small. Repeating the experiment in these conditions (and counting for a suitably extended time) we get exactly the same result as before.

A conclusion commonly drawn in orthodox quantum mechanics is this:

> We must conclude that when both holes are open, it is *not true* that the particle goes through one hole or the other. For if it had to go through one or the other, we could classify all the arrivals at [D] into two disjoint classes, namely, those arriving through hole [B] and those arriving through hole [C]; and the frequency [N(B and C)] of arrival at [D] would surely be the sum of the frequency [N(B)] of particles coming through hole B and the frequency [N(C)] of those coming through hole [C].[11]

This obviously presupposes, in our terms, that the motion of a particle can always be derived from an OCF specific to itself, as is true for a classical particle. As we have noted, once random disturbances are introduced, there is no way, in a variational treatment, to avoid considering the particle as a member of an ensemble, whether or not members of the ensemble are present simultaneously. Then the motion of the particle is no longer independent of the ensemble to which it belongs.

In our treatment, the behaviour of the electrons follows in a natural way if we assume that they belong to an ensemble in which it is not possible to say which hole a particle passed through. The discussion given for the fishing fleet and the sandbanks can be transferred directly to the ensemble of complex images. Then the behaviour of the ensemble of particles can be related to the behaviour of the complex images.

If on the other hand, we could say of each particle arriving at D that it had actually passed through hole B, or through hole C, we should expect to find that N(B and C) was the sum of N(B) and N(C). If we detect the electrons as they pass through one or other of the holes, this is indeed what we find when they arrive subsequently at D.

The objection may be raised, how can a particle determine the optimal path to satisfy a variational principle for an ensemble, but the question is misconceived. We may as well ask, in the orthodox theory, how a particle can find its way from one side of the screen to the other without passing through either hole. As always, the phenomena belong to nature, but the explanations are ours. We need not suppose that electrons have any more knowledge of control theory than of the axioms of orthodox quantum mechanics. To satisfy a variational principle is to implement the policy which can be obtained from it, and our aim has been to make this policy agree with the standard prescriptions of quantum theory.

The variational principle we have described can be used to derive a wide range of results[9] which have been established in the orthodox theory. To show that the two approaches are equivalent, we ought to show that this is true for all results (or at least all known results) in the orthodox theory. We also ought to show that all results in the variational theory can be derived from the orthodox theory. The first statement, if it is true, demands far more effort for its establishment than has yet been given: orthodox quantum mechanics has no doubt received many thousands of times more attention than any variational alternative. The second, besides the effort needed, is hampered by the present state of orthodox quantum mechanics. Though this is highly developed, it is by no means in such a clear and well-formulated state as classical mechanics.

8. *Is orthodox quantum mechanics causal?*

We have spoken of the orthodox theory of quantum mechanics as being causal, and have contrasted it in this respect with a theory based on a variational principle. Some would deny that quantum mechanics is causal, on the grounds that physical variables in this theory are random, and cannot be predicted. In this understanding of the term, 'causal' implies 'completely deterministic'.

This is a matter of definition, like the use of 'causal' to mean simply 'non-anticipative' which is sometimes found. As a matter of common usage, 'causal' is not generally restricted to fully deterministic systems. In ordinary circumstances, the height of pedestrians taken in sequence as they walk along the pavement is a random variable. We cannot predict the height of the next pedestrian from any measurements we have taken in the past.

Nevertheless, we can predict the average height of the next thousand pedestrians with good accuracy. We can also, given previous experimental data, predict the probability distribution of heights: what proportion of the thousand will be taller than 170 cm and not taller than 175 cm, etc.

We can also make assertions such as 'the average height of the population has increased by 2 cm in the past century'. We can moreover look for causes of this increase, and can assert for example, that it is the effect of better nutrition. The fact that variables are random does not prevent us from postulating causal relations for certain aspects of their behaviour.

It is the same with quantum mechanics. The theory contains random quantities, but it also contains quantities which are not random. The wave function is non-random. It obeys a differential equation and evolves deterministically. From it we can predict for example, the mean velocity which we shall obtain for particles in a particular experiment, and the probability distribution of velocities. It is predictions of this kind which render the theory useful.

The contrast we have drawn is between predictions made, as in the orthodox theory, by causal relations, expressed usually as differential equations, and predictions based on a purpose expressed as a variational principle. From the variational principle we can generate the policy, which is the set of causal relations that implement the purpose expressed by the variational principle. This policy will agree with the orthodox theory. Alternatively, we can argue directly from the variational principle, as we did when considering the interference of electrons passing through two holes in a screen.

9. *Information as a cause*

We have noted on several occasions that a variational principle has to be satisfied under the constraint that only some kinds of information are available. One constraint of this type which we have used everywhere is that some information about the past is available (and about the present as a limiting case) but no information about the future. Our policies, obtained subject to this restriction, are non-anticipative.

The constraint is appropriate in all the cases we have considered, but it is imposed externally, rather than inherent in the theory. For example, consider the problem of removing noise (that is, random disturbances) from a video picture transmitted from a satellite in space. We can pose this problem in a number of forms, including the following.

i. We can ask, during the transmission of the picture, for the best result available from the information transmitted so far. The noise will be removed by some form of averaging: a random fluctuation in transmission which affects one point on a scan line will be unlikely to affect neighbouring points on the same line to the same extent. Still less will it be likely to affect neighbouring points in adjacent scan lines. Knowing statistical properties of the noise, and of the desired signal, we can set up a computer program which compares adjacent points and minimizes some 'cost' expressing the difference between the received signal and the transmitted signal, averaged over a large number of transmissions: this objective can be expressed as a variational principle.

Doing this during the course of transmission, we shall have available the previous scan lines, and part of the present scan line. We shall have no information about the part of the picture still to be received. Our computer algorithm will be designed to do the best possible with this information, and will be non-anticipative.

ii. We can record the whole transmission of the picture before we start processing it to remove the noise. Then in considering each point in the picture, the computer will have available information from all adjacent picture points, whether they were transmitted before or after the one considered. A different algorithm will be used, and it will produce a better result.

This procedure is non-anticipative in the essential sense—we cannot start processing the picture until the whole of it has been received. But as compared with the process in (i) we can now use information which was then in the future, and unavailable. The penalty we pay for the improved result is the delay which occurs before it is available.

iii. Because local points have most influence in the computer algorithm, we can produce nearly as good a result as in (ii) with a smaller delay. We can, for example, process the points in one scan line after the next two scan lines have been received.

In all cases, we are subject to the constraint of not knowing the future. When we derive the computer algorithm, which implements the policy satisfying the variational principle, we include this constraint in the mathematical formulation. The form in which we do so will depend upon the particular situation we are considering. These comments clarify the relationship of non-anticipation to variational principles, and to the causal relations expressed by a policy obtained from the variational principle.

What has been said so far relates to information expressed in numerical form. This kind of information is equally at home in variational treatments of physics and in causal treatments. In Section 7, we considered a second kind of information: did a particle pass through hole B or hole C? As we saw, the solution to the variational problem depends quantitatively upon whether the answer to this question is assumed to be known or not known.

To incorporate such information in a causal treatment, by a quantitative mechanism divorced from any underlying variational principle, does not seem to be possible. This is at the heart of the difficulty experienced by orthodox quantum mechanics in dealing with the interference of electrons passing through holes in a screen.

That the information does quantitatively affect the outcome of the physical experiment described in Section 7 is certainly true. And the question is not whether we actually know this information, but whether it could have been known:

> For example, suppose that information about the alternatives [whether each electron passed through B or through C] is available (or could be made available without altering the result), but this information is not used [sc. by the physicist]. Nevertheless, in this case a sum of probabilities [N(B) plus N(C)] must be carried out.[12]

What we observe, that is whether *N*[B and C] is or is not equal to the sum of *N*[B] and *N*[C], thus depends upon whether information is available about which hole was used by each electron. It does not matter whether we use this information, nor even whether we know it. Provided that it is available to be known from the experiment without any change in the latter (but not otherwise) we shall find that *N*[B and C] is the sum of *N*[B] and *N*[C]. The information therefore acts as a cause of this equality.

This is what we predict from our variational principle, just as we did for the fishing fleet. There, with information available about which channel each boat had passed through, the purpose of minimizing the fuel consumed could be fulfilled more effectively than when this information was lacking. The policy, and hence the behaviour of the boats, was modified by the availability or otherwise of the information.

Lacking any mechanism by which information can act as a cause, orthodox quantum mechanics relies upon an axiomatic statement. Two different situations are envisaged—the information is available or it is not available—and the outcome in the first situation differs from that in the second. The physicist proposes one outcome or the other according to which situation he judges to exist, thereby importing an awkwardly subjective element into the theory. At the same time, as we have seen, it is denied (when the information is not available) that electrons actually pass either through one hole or the other.

These comments about the conceptual difficulties faced by the orthodox (causal) theory do not affect the question whether the variational and the causal approaches to quantum mechanics are equivalent. Axiomatically defined procedures in the orthodox theory will still be equivalent to a variational principle if they lead to the same result.

10. *Subordinate purposes*

Let us return now to the simple case considered in Chapter 2: a ship is sailing under automatic control, in still water, in a way which minimizes the amount of fuel used. From what has been said in this chapter and in Chapter 3, we can propose a different and deeper variational principle to describe the ship with its engine, rudder and control system. We regard all of this as a physical system, and we can then apply Hamilton's principle, and the principles described in this chapter, to give a variational treatment at a detailed level.

We can apply Hamilton's principle to describe the motion of parts of the engine. The chemical reactions of combustion in the engine can ultimately be derived from quantum mechanics, expressed in a variational form. The mechanical and electronic actions of the control system can be expressed similarly. Admittedly we cannot actually carry out this programme. Some essential parts of the variational treatment of quantum mechanics are missing. Even if the theory were complete, we could not carry out the derivation in detail from the variational principles, any more than a causal description in terms of molecular forces and motions could be given.

The point we wish to discuss, however, is a conceptual one. Grant that a physical description can be given, resting on a variational principle. This determines the motion of the ship by reference to a purpose. But the motion of the ship is also determined by a different purpose built into the controller, namely that the fuel consumption shall be minimized. How can one and the same system fulfil two different purposes simultaneously?

One of the two purposes, the physical one expressed by Hamilton's principle and its extensions, is certainly more detailed than the other, engineering, purpose described by the minimization of fuel. But this difference does not remove the difficulty. The physical purpose will generate a description of many details which cannot be obtained from the engineering purpose, but it will also generate those details which do arise from the engineering purpose. From the physical purpose, for example, we can obtain the course of the vessel, its speed, and its fuel consumption.

It is clear that the two purposes must bear some special relationship to each other if they are to be compatible in this way. To understand how the relationship arises, let us consider how the control engineer approaches the task of incorporating an engineering purpose. He sees this as a problem of designing the control system (an automatic navigation system, based perhaps on a computer) together with the instruments which will supply the necessary information, and the actuators which will allow the controller to influence the behaviour of the ship. The instruments, for example, could measure the speed and heading of the vessel, as in Fig. 2.11, and the actuators could affect the throttle and the rudder.

Then, if he uses dynamic programming to design the controller, he will start by writing down an equation which shows how the fuel consumption depends upon the working conditions of the ship and the engine. The consumption will depend, for example, on how wide the throttle is opened, and how fast the engine is running. Then the variables which can be manipulated—the throttle setting and rudder position—have to be determined by the dynamic programming argument so as to minimize the total consumption of fuel.

The problem is not yet fully described, however. For a given throttle setting, the engine will turn faster when the ship is moving than when it is stationary, because the propeller will impose a smaller resistance when the ship is in motion. The resistance to the ship's motion at a given speed, and hence the fuel consumption, will be greater if the rudder has been moved from its central position. These and other complicated relations have to be described mathematically and written down. When this has been done, the only things left undetermined are the throttle setting and rudder position. Given these, everything else can be calculated, including the total amount of fuel used. We are then in a position to solve the problem, using dynamic programming.

We have, incidentally, refined the problem beyond the description given in Chapter 2. There we assumed that the fuel consumption depended upon the square of the vessel's speed. This will be true at any steady speed with our present description, but we have also allowed for minor effects such as the fuel used to accelerate the vessel.

Now the details we have just described are a part of the physical description of the engine and ship—that part of the physical description which is needed to implement the engineering purpose. We are saying, in effect, that the ship and its engine behave in such and such a way when we alter the throttle and rudder settings. Given this, how should we manipulate these settings, in order to use the minimum amount of fuel? We look for the answer in terms of a policy, showing how the settings should depend upon the measured variables. The latter are the heading and speed of the ship, from which in still water we can obtain its position by dead reckoning.

It will now be seen why the engineering purpose is consistent with the physical purpose—we wrote the significant part of the latter into the description of the problem which defines the former. To do so, we had to express the physical purpose (that is, Hamilton's principle with any necessary extensions) as a set of causal relations: those causal relations which can be deduced from the physical purpose. Incorporating these causal relations in the engineering purpose, we can deduce the causal relations which define the control system, and which allow us to build it, Fig. 5.9. The passage in this figure from purposes to the equivalent causal relations (A to B and C to D) can be made by dynamic programming.

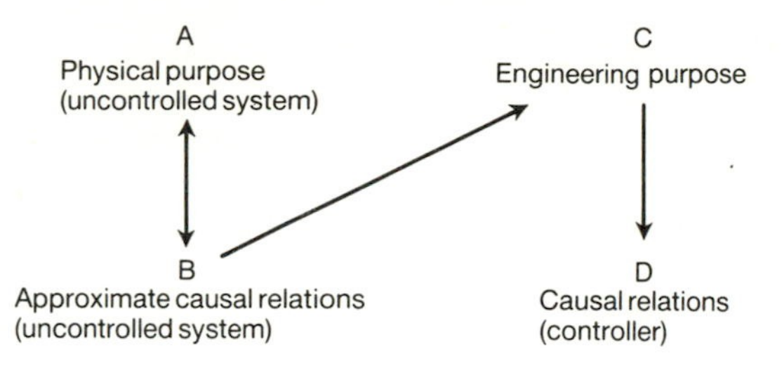

Fig. 5.9 Designing a control system.
At A we have a physical purpose for the uncontrolled system expressed by Hamilton's principle (with possible extensions). By dynamic programming we can deduce the equivalent causal relations B. Together with the engineering purpose C these define (by dynamic programming) the causal relations D of the control system, which allow us to build the controller.

The route shown in Fig. 5.9 is an indirect one for two reasons. First, we do not have mathematical techniques for using Hamilton's principle or its extensions (A) as a constraint on the engineering purpose (C), but first have to derive from it causal relations (B) which are a sufficiently accurate description for our purposes. Secondly, we cannot build a controller directly from a definition of the engineering purpose (C), but first have to deduce the equivalent causal relations (D). In both cases it seems likely that the universal development of science and engineering in causal terms, and the universal rejection of purpose as a valid alternative, are at least partly responsible.

From the approximate causal relations B of the uncontrolled physical system, and those D of the controller, we can construct the controlled system (the original physical system connected to the controller) shown as E in Fig. 5.10. E has a fundamental description F in causal terms, which in

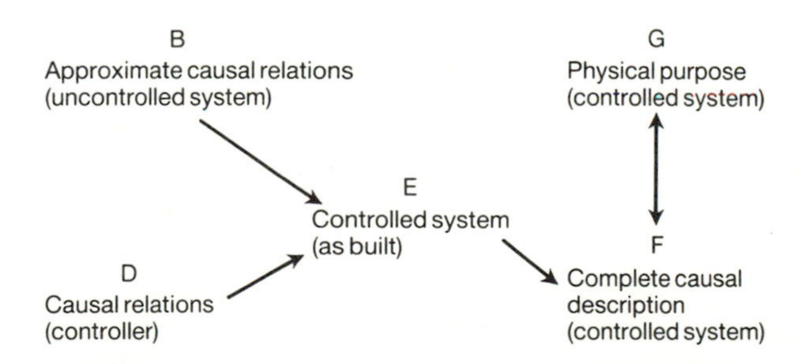

Fig. 5.10 The causal description B of the uncontrolled system, together with the causal description D of the controller, give the causal description E of the controlled system. There is a more complete description F of the controlled system in terms of quantum mechanics. This can be derived from a variational principle G expressing a fundamental purpose.

principle can be obtained from quantum mechanics. This description F expresses the policy which fulfils a purpose described by a variational principle G. Then both C and G express purposes which are fulfilled by the controlled system. We shall say that the engineering purpose C is subordinate to the purpose G defined by an extended form of Hamilton's principle.

Interpreting what has been said in relation to the ship, B is an engineering description of engine, ship, rudder etc., in terms of differential and algebraic equations. C is the purpose of going from one given point to another given point with the least consumption of fuel. D is a set of differential and algebraic equations which describe the controller needed to accomplish C subject to B. E is the complete engineering system of ship and controller which automatically satisfies C. This system, as built, can be described in principle by a detailed physical description F, still in causal terms. Then there exists a physical variational principle G expressing a purpose from which F can be derived. We say that the purpose C is subordinate to the purpose G.

A second example, which does not depend upon control engineering, may help to make the idea clearer. Suppose that the components of a gearbox are laid out separately on a bench. Each of them, when it is lifted and rotated, conforms to Hamilton's principle. When the parts are assembled, their motions are severely constrained, so that only a very few of the motions which could previously occur are possible.

We can again write down Hamilton's principle, for the assembled gearbox, and in doing so we need take into account only the possible motions of the parts: this is one of the major practical advantages of the principle. The permitted motions when the parts are assembled are certainly among those which were potentially possible when the parts were separate. The purpose expressed by Hamilton's principle for the assembled parts is then subordinate to the more complicated and all-embracing purpose for the separate parts.

We shall use the same term, 'subordinate purpose', to describe a wide range of variational principles which have been developed for the description of physical systems.[13] For example, the flow of heat in a solid can be described at the engineering level by a variational principle, which can be used effectively for purposes of calculation. But at a deeper level, the heat energy can be regarded as the energy of elastic waves passing through the solid, and these can be treated by quantum mechanics.[14] Assuming that the latter treatment can be brought within the scope of the variational principle described earlier in this chapter, we shall say that the engineering principle is subordinate to the quantum-mechanical principle.

References

1. Max Born (1965). *Einstein's theory of relativity*, pp. 225–372, Dover.
2. Cornelius Lanczos (1970). *The variational principles of mechanics*, pp. 291–340, University of Toronto Press.
3. Tullio Levi-Civita (1929). *The absolute differential calculus*, Blackie; C. E. Weatherburn (1966). *Riemannian geometry and the tensor calculus*, Cambridge University Press.
4. Leonard A. Schiff (1968). *Quantum mechanics*, McGraw-Hill; Bernard d'Espagnat (1971). *Conceptual foundations of quantum mechanics*, W. A. Benjamin.
5. Wolfgang Yourgrau and Stanley Mandelstam (1968). *Variational principles in dynamics and quantum theory* (3rd edition) p. 177, Pitman.
6. Harold J. Kushner (1967). *Stochastic stability and control*, Academic Press; Ludwig Arnold (1974). *Stochastic differential equations*, Wiley.
7. E. Nelson (1967). *Dynamical theories of Brownian motion*, Princeton University Press.
8. Francesco Guerra (1981). Structural aspects of stochastic mechanics and stochastic field theory, *Physics Reports*, vol. 77, pp. 263–312, North-Holland; S. K. Mitter (1981). Non-linear filtering and stochastic mechanics. In M. Hazewinkel and J. C. Willems (editors), *Stochastic systems: the mathematics of filtering and identification and applications*, pp. 479–503, Reidel; Francesco Guerra and Rossanna Marra (1984). Discrete variational principles and quantum mechanics, *Physical Review* D, vol. 29, pp. 1647–55; Akira Ohsumi (1989). Derivation of the nonlinear Schrödinger equation from stochastic optimal control, *Int. J. Control*, vol. 49, pp. 841–9.
9. H. H. Rosenbrock (1985). A variational principle for quantum mechanics, *Physics Letters*, vol. 110A, pp. 343–6; (1986). On wave/particle duality, *Physics Letters*, vol. 114A, pp. 1–2; (1986). Three approaches to quantum mechanics, *Physics Letters*, vol. 114A, pp. 63–4; (1986). The quantum-mechanical probability density, *Physics Letters*, vol. 116, pp. 410–12; (1987). Outline for a variational development of quantum mechanics, *MTNS Conference*, Pheonix, Arizona, 15–19 June.
10. R. P. Feynman and A. R. Hibbs (1965). *Quantum mechanics and path integrals*, pp. 2–5, McGraw-Hill.
11. Reference 10, p. 6.
12. Reference 10, p. 14.
13. Maurice A. Biot (1970). *Variational principles in heat transfer: a unified Lagrangian analysis of dissipative phenomena*, Oxford, Clarendon Press; Saul T. Epstein (1974). *The variational method in quantum chemistry*, Academic Press; Karel Rektorys (1977). *Variational methods in mathematics, science and engineering*, D. Reidel.
14. H. A. Wilson (1944). *Modern physics*, pp. 79–82, Blackie.

6 The purposive myth

How it would seem if we believed it

1. *Basis of the discussion*

In Chapters 3 and 5 an outline has been given of an alternative description of nature, based upon purpose rather than causality, but equivalent to a causal description in the sense defined in Chapter 4. According to that definition, neither myth expresses the 'reality' of the world, that task being reserved to the equivalence class containing both myths.

But the present situation is that by an act of faith the causal myth is almost universally believed to be the one 'true' and 'real' account of nature. It is not regarded as a human construct, intended to explain the world and give us power to control it. Instead, the myth is regarded as an accurate description of causal machinery, by which the world functions and by which all events in it are produced. This belief permeates our science and our technology, and our whole outlook upon the world and our behaviour in it. To demonstrate how deep this influence is, and how pervasive, the simplest course is to give a view of the world as it would appear if, by a different act of faith, we accepted the purposive myth with the same tenacious belief in its truth as we now give to the alternative.

This is what will be done, so far as it is possible. The scientific and technical aspects of the account are straightforward, limited only by the inadequacies of any one person's grasp of the whole area to be covered. The wider consequences for our beliefs and behaviour are more difficult to predict, and demand an extrapolation which must tax the imagination.

2. *Purposive physics*

Our ordinary view of the world is based upon our experience of objects having an appreciable size—of at least the size which we can see with the naked eye. When we descend to much smaller particles, and try to explain their behaviour, we find that it departs very markedly from what we should expect. As an example, let us compare the behaviour of a stone, such as we might pick up in our hand, and of an electron.

The purpose fulfilled by a stone is specific to itself, so that we can talk about the behaviour of a single stone: when we pick it up and throw it how do we affect its purpose? To define the purpose we need to know two things, the potential energy of the stone and its kinetic energy. The potential energy increases directly as the height, while the kinetic energy is proportional to the square of the speed. Then the stone moves in such a way that a certain quantity called the 'action' is a minimum. The contribution to the 'action' at each instant is the kinetic energy less the potential energy, and the stone moves, according to Hamilton's principle, so that the sum of all the contributions during its motion is a minimum. (Here and later we talk about a minimum for simplicity: more generally we should talk of a stationary value.)

When we throw the stone, we impart to it a certain initial velocity, that is, an initial speed and direction, and it has a certain initial position. Thereafter, the stone moves so that the 'action', taken over its path, is a minimum. All paths which the stone might follow, with different initial positions and velocities, make the 'action' a minimum. The initial velocity which we actually impart to it serves, with the initial position, to select one from the multitude of paths which the stone might follow. The purpose is constant—minimization of the 'action'—but how it is expressed in the motion of the stone depends upon the particular circumstances. These circumstances are the initial position and velocity, the way the potential energy depends on position, and the way the kinetic energy depends upon the velocity.

For an electron, on the other hand, the situation is different. It fulfils a purpose, but we cannot define its purpose except as a member of a group, or ensemble. This has nothing to do with the electrical or other forces between electrons, but is true even when the electron moves as a single individual. It still has to be considered as a member of an ensemble of electrons, each moving individually under the same conditions.

The reason for this is that the motion of the electrons is derived from the motion of 'complex images', which are subject to random disturbances. The purpose which the complex images fulfil can therefore only be described in an average sense, taken over the whole ensemble of images. All of the complex images then follow a policy which makes the 'action', taken over this ensemble, a minimum. By considering the ensemble, we average out the random disturbances which affect each complex image. The policy makes the 'action', for the ensemble as a whole, as small as possible in the face of the random disturbances. Then the motion of the ensemble of electrons is derived from the motion of the ensemble of complex images.

For non-interacting particles, we consider an ensemble consisting simply of a collection of systems each having a single particle. If we need to consider a number of particles which interact with each other, for example interacting through electrical forces, then the ensemble will consist of a collection of systems, each containing the same number of particles interacting in the same way. It is to this ensemble, through the mediation of the complex images, that the variational principle will apply.

Electrons, therefore, influence one another not only by the physical forces which they may exert, for example because of their charge, but also by belonging to the same ensemble. That is to say, the information that they belong to one ensemble rather than another will change the way they behave. Figuratively speaking, we may say that the purpose of electrons is a social one, where the purpose of the stone is individual.

All small enough particles behave in this way, whether they are charged or not. Indeed, all particles, whether large or small, do so, but the 'social influence' becomes less the larger they are. This is because the random disturbances, which compel us to average over an ensemble, become smaller as the mass of the particle increases. So if the particle is massive enough, all members of the ensemble to which it belongs behave in substantially the same way. The random disturbances become inappreciable, and the average taken over the ensemble is the same as the value for one particle. We can apply what has just been said directly to the motion of a stone by regarding the stone as one 'particle', having a large mass and belonging to an ensemble of stones.

More accurately (though it is usually unnecessary) we ought to regard the stone as a collection of subatomic particles—electrons, protons and neutrons. These are bound together by electrical and other interactions, and retain (to a high degree of certainty) the essential features of their relative locations, that is, of belonging to the stone. They will have vibrational energy corresponding to the temperature, and some of them may move with relative freedom from point to point within the stone, but they will remain recognizable as belonging to the stone.

To make a quantum-mechanical analysis of this system of interacting particles, we consider it as a member of an ensemble of systems. Certain properties of the members of this ensemble will differ significantly from one member to another: for example the vibrational behaviour. Other properties, such as the motion of the centre of mass, will differ little from one member to another of the ensemble, and will behave very nearly according to the classical prediction. These properties, therefore, can be derived from Hamilton's principle.

3. *Classical and quantum-mechanical worlds*

The purpose of the stone, expressed by Hamilton's principle, belongs to classical mechanics. It is an example of a 'subordinate purpose' as described in Chapter 5. That is to say, it can be deduced from the more detailed purpose of the subatomic particles of which the stone is composed, and as an approximation is consistent with it, but is a simplification giving only a part of the information which can be deduced from the underlying quantum-mechanical purpose.

If we look in detail at the quantum-mechanical description, we shall find that the identity of the stone, which is unambiguous in the classical description, loses its sharp definition. We cannot say with certainty that a subatomic particle belongs to the stone. We can say so with an extraordinarily high degree of probability, but because of the random disturbances it will always be possible for particles which belonged to the stone (that is, were contained in a certain envelope around it) to leave it, and for others to join it.

The sharp classical distinction between the object which we study and its environment can no longer be made. We are therefore constrained to enlarge the ensemble which we consider, so that it includes some part of the 'environment': that part of it which we cannot distinguish unequivocally from the object we wish to study. In doing so, we change the purpose which the object under study fulfils, through the phenomenon of interference. And once started on this process of enlarging the ensemble, there is no point, in principle, at which we can stop.

This fact leads to a difference between the way in which we conceptually construct the classical world, and the way in which we must construct the quantum-mechanical world. In the former, we start as in Newtonian mechanics with an isolated particle, which when subject to no external influence fulfils its purpose by travelling in a straight line at constant speed. Then we consider the effect on it of external influences expressed by forces. Gradually, in this way, we build up more and more complicated representations, starting from the simplest basis of an isolated particle following a unique path.

This approach is not possible in quantum mechanics. An isolated particle there does not pursue a unique path, because of the random disturbances. The purpose expressed by the particle only has meaning for an ensemble of particles. But because of the random disturbances, we are generally not in a position to say with absolute certainty whether a particle belongs to a given ensemble or not. And this uncertainty, in principle, changes the motion of the particles resulting from the purpose they express, though the extent of this change may in many circumstances be negligibly small.

The fact is that in quantum mechanics we cannot unambiguously distinguish between an ensemble of particles which we wish to study, and the rest of the world. The difficulty goes beyond the difficulty in the classical theory: that when we consider an isolated particle there must always be some influence on it by the external world, whether gravitational or electrical, and however small this influence may be made. Within the classical theory this difficulty exists, but it leaves unambiguous the object—the particle—which we wish to study. We are then able to estimate the external influences, and assess their effect.

For a quantum-mechanical system, the same problem exists, and can be dealt with in similar ways, but there is a more insidious difficulty. Because of the random disturbances, particles 'belonging to the external world' and particles 'belonging to our ensemble' will always be to some extent mixed together and confused. Then even if they exert no 'physical influence' through forces between particles of the ensemble, these external particles will change the behaviour of the ensemble.

If, for example, we are studying the motion of electrons, we have to recognize that the environment, the experimental apparatus, also contains electrons. We cannot distinguish these with certainty from the ones we wish to study, so we must extend our ensemble to include them. But then the behaviour of the electrons under study will be changed.

The reason was given in Chapter 5, and was illustrated there by the interference of electrons passing through two holes in a screen. In that experiment, think of electrons which pass through hole B as the ensemble we wish to study. Think of the electrons which pass through hole C, perhaps quite far away, as belonging to the external world. When we cannot say which hole a particle has passed through, we have to treat all of them as belonging to one ensemble. Then the members of this ensemble which pass through B behave in a way which is different (though perhaps only slightly different) from the behaviour of the ensemble we wish to study: namely those particles which pass through B in the absence of those which pass through C. The problem is compounded by the fact that some interaction with the external world is essential if we wish to observe the behaviour of the ensemble in which we are interested.

There is thus the following conceptual difficulty. If we try to construct the world 'from the bottom up', in quantum mechanics, we can never define unambiguously the object (an ensemble) which we wish to discuss. We may be able to make this definition with sufficient certainty for most practical purposes, but conceptually the barrier is absolute. We are therefore driven to adopt a different approach, the opposite one: of constructing the world 'from the top down'.

The set of all particles which exist, the whole universe, can be regarded as a member of an ensemble of universes, and for this we can define a purpose, and in principle can evaluate a policy. We call these the universal purpose and the universal policy. Then below the universal purpose is a complicated network of subordinate purposes. The motion of the planets in the solar system is very accurately approximated by the policy derived from Hamilton's principle, with a small correction for Mercury from general relativity.

A great many of the features of the world with which we are familiar on earth are equally derivable from Hamilton's principle. On a smaller scale there are different subordinate purposes, expressed in quantum mechanics, for systems which we regard to a sufficiently close approximation as isolated from their environment. Subordinate to these again are purposes such as the variational principle which can be used to describe the flow of heat.[1]

All of these subordinate purposes give approximations (often extremely close) to results which, in principle, can be derived from the universal purpose. We say 'in principle', because it will certainly remain impossible to deduce every result in which we are interested in this way. The caveat is of the same kind as must be used within the alternative, causal, myth. There, in principle, everything can be explained by putting together more and more complicated assemblies of simple particles, because what we have described here as an approximation valid for classical systems is assumed there to be true universally. But, even if true universally, it is accepted that the proposed explanation can never be carried out in full detail.

4. *Biological purposes*

Living organisms show behaviour which we recognize in an everyday sense as being purposive. The green protozoan Euglena, which needs light for photosynthesis, is equipped with a flagellum by means of which it swims through the water. In a jar exposed to light the organisms will congregate in the part where the light is strongest. Or if they are viewed under the microscope by dark-field illumination, they will crowd around the cone of light, continually darting through it.

An explanation of this behaviour in causal terms, that is in terms of the policy, follows the same lines as are used in explaining the behaviour of a heat-seeking missile. There is a sensor, the stigma, which responds to the intensity of light. There is an actuator, the flagellum, which imparts a velocity to the organism in response to information reaching it. Between these two is a controller, difficult to identify, which receives information about light intensity and generates information to command the actuator. This is a closed-loop feedback system, because motion of the organism changes the light intensity which it senses and responds to.

Given details of the policy, we can seek to construct a purpose which gives rise to it. We look, that is, for a variational principle, which gives rise to the observed behaviour as the policy which will satisfy it. Within the orthodox framework of the causal myth, no attempt is made to do this, so that no appropriate principle will be found in the literature. The question then arises, does there actually exist a variational principle giving rise to the behaviour we observe?

Policies arising from particular kinds of variational principle have quite well-defined properties, so that we may be able to say with certainty that a given policy could not arise from a given type of variational principle. But we cannot go further and say of a policy that it could not possibly arise from any variational principle whatever, because the range of these principles is limited only by our ingenuity. On the other hand, it may be equally impossible to show directly, until after we have actually found it, that an appropriate variational principle does exist for a given policy.

Fortunately the question, in principle, is already answered. The universal purpose applies to the whole universe. To the extent that we can isolate and identify a system of particles constituting Euglena and its environment, there will be a subordinate purpose which, as a very close approximation, governs its whole behaviour. Subordinate to this again will be a purpose from which, with further approximation, the light-seeking behaviour of Euglena can be derived. A derivation of the appropriate variational principle in this way is readily admitted to be beyond our abilities. The result might also be excessively complicated. Nevertheless, the existence of the desired principle is not in doubt.

This discussion could be extended to a multitude of different aspects of the behaviour of living organisms, both less and more complicated than Euglena. In the human body, for example, there are control systems of great complexity which govern body temperature, or the circulation of the blood, or other variables.[2] There are feedback mechanisms governing hand–eye co-ordination. There are also mental processes governing more general types of behaviour. Nothing further need be said of any of these, because they can all be brought within the framework already suggested.

The origin of life and the progress of evolution can equally be brought within an extended form of the same argument. There is a causal description of this process, partly known and partly guessed at. It is based, in its present form, upon the replication of organic molecules by a mechanism which can be fairly completely described.[3] Molecules of DNA code for the structure of proteins, and are passed on unchanged from generation to generation of a species. Unchanged, that is, apart from rearrangements during sexual reproduction, and the occurrence of rare but seminally important errors. The results of these rearrangements, and of the errors, in turn replicate themselves, but with varying success. Those which are most successful come to dominate their populations.

This outline, originating with Darwin and successively refined, will no doubt be subject to still further refinement and correction. Whatever may be its final and accurate form, we can say at once that it is the policy which expresses some purpose subordinate to the universal purpose. Based as it is upon quantum mechanics, there is an essential element of randomness in the latter, just as there is in the description of the process by which the policy operates. But the purpose is fulfilled to the highest degree possible in the face of this randomness.

The real difficulty in this description of evolution as a purpose, operating in the face of random disturbances, lies in identifying the appropriate system, the appropriate collection of particles, to which it should be applied. We need to specify a subset of the universe to which we can assign, through the ensemble to which it belongs, a subordinate purpose. The subset must be clearly identifiable, and nearly independent of what we exclude.

Individual organisms do not qualify, nor individual species. One species is a part of the environment for another: the evolution of a predator influences the evolution of its prey and vice versa. We have, at least, to consider the whole biosphere, that part of the earth's surface and atmosphere inhabited by living beings. This in turn has been subject to influences from the solar system—light and heat from the Sun, tides caused by the Moon, and so on. Errors in replication of DNA have been caused not only by local quantum effects, but also by cosmic rays emanating from outside the solar system.

Whether evolution can be regarded as a phenomenon belonging to something less than the whole universe is therefore an open question. The difficulty, as mentioned earlier, does not lie simply in the influence exerted through physical effects such as gravity and electrical or magnetic forces. In addition, adding new particles of a given type to an existing population changes the physical behaviour of all the particles concerned.

5. *Technological systems*

At the opposite extreme from biological systems are those man-made systems which incorporate a human purpose. The origin of this human purpose must be found in the account given in the previous section. The way in which the purpose is incorporated in a machine has been described in Chapter 2. What is the status of a purpose which has been incoporated in this way?

In the causal myth, the purpose effectively disappears. The machine is described causally by the policy derived from the purpose, and the purpose is either denied or ignored. This is the substance of the statement that 'man is a machine', which carries over to man the same denial or ignoring of purpose. In the purposive myth, on the other hand, a purpose which has been incorporated by human action in a machine retains its validity as a purpose. It can be translated into a policy to give a causal description, but equally it can be derived as a subordinate purpose from more fundamental purposes.

The machine will have a purposive description in terms of Hamilton's principle, for the mechanical components, which is subordinate to a description in terms of a quantum-mechanical principle. Electronic devices can equally be described by the quantum-mechanical principle. Subordinate to these descriptions will be the description of a machine in terms of the human purpose which has been incorporated in it. That is to say, this human purpose now exists as a simplified description of behaviour, which behaviour can be deduced from the more fundamental, quantum-mechanical, purpose of the machine.

If, therefore, we describe the machine at a fundamental, quantum-mechanical, level by means of its purpose, this fundamental purpose now has, as a subordinate purpose resulting from it, the human purpose we incorporated. We have, by incorporating the human purpose, modified the fundamental purpose of the machine. This has been achieved by the indirect process described in Section 10 of Chapter 5, because we have no present technique for directly modifying a fundamental purpose in this way.

A first consequence of this account is that moral judgements of human purposes do not cease to be valid when those purposes are incorporated in machines. We may, for example, condemn a human purpose. When this purpose is incorporated in a machine, according to the causal myth, the purpose disappears. We are left with a device which follows certain causal laws, but has no purpose, and so is not subject to moral judgement. Within the purposive myth, the human purpose persists as a subordinate purpose of the machine, as it was earlier a subordinate purpose of a human being, and the same moral judgement is appropriate. Our instinctive moral repulsion from a thermonuclear missile is justified in the purposive myth, whereas it has no force or validity in the causal myth.

6. *Man and computer*

A second consequence of our account bears on a conclusion which is often drawn from the alternative, causal, myth. Briefly stated, this is that 'man is a machine'. Therefore anything which a man can do can be done by a machine. In particular, it can be done by the universal machine, the computer, if this is provided with appropriate sensors and effectors. This argument has been put forward particularly by exponents of artificial intelligence, and is very difficult to counter within the framework of the causal myth.[4] The implicit conclusion which is suggested is that a computer can be programmed so that (with appropriate sensors and effectors) it is fully equivalent to a man.

If we believe the purposive myth, human behaviour takes on a different aspect. In principle, science can give a description of the human body based on quantum mechanics. An immensely numerous collection of fundamental particles forms a system which we can identify, with more or less certainty, as constituting the body at a given time. These particles move and interact in such a way as to fulfil a fundamental purpose, described by a variational principle. The description, because of its complexity, will always be inaccessible to us in detail, just as a causal description in terms of the policy derived from the purpose is inaccessible. Nevertheless, the fundamental purpose exists according to the act of faith by which we accept the purposive myth.

Subordinate to the fundamental purpose are many purposes which describe particular aspects of our activity. Some of these subordinate purposes express bodily functions: the repair of damage, the maintenance of structure, the circulation of the blood. Others resemble closely the kind of control system which we can build into machines: for example the mechanism by which body temperature is maintained close to a fixed value.

Still other subordinate purposes describe common actions such as walking or running, or other activities which we have learned: driving a car, playing squash, or using a typewriter. All of these we do, once they are familiar, without conscious thought. The purpose is translated directly into action, without intermediate analysis, just as the earth fulfils Hamilton's principle in its passage around the sun without ever formulating or solving the differential equations which express the policy.

Some parts of the purpose are present in consciousness. In driving a car, for example, we shall concentrate our attention when overtaking. We shall have a purpose, which we implement, but if we are experienced drivers we shall do so without any prior translation of the purpose into a policy. We shall change gear, and accelerate, and steer, without conscious thought and often without direct awareness.

The simple and traditional kinds of human work have exactly this character of directly implementing a purpose. When we use a tool, it participates in our purpose. Following Polanyi,[5] for example, we may say that when we use a hammer, we 'feel' directly the impact of the head upon the nail. We feel whether the blow is true, whether the nail is driven, whether it is solid, or whether it bends. The purpose of an independently existing hammer, expressed by Hamilton's principle, has been subsumed into our own purpose. The hammer becomes an extension of ourselves.

By the agency of the hammer, our purpose is converted directly into action, without first being translated into the policy. We could, with sufficient labour, and as an intellectual explanation of our actions, deduce the policy which we are implementing with the aid of the hammer. The policy would give a causal picture, showing how the impact of the hammer head results in a reaction between the handle and the palm of our hand, and how by a process we cannot follow in detail, this reaction is converted by the brain into the perception of a blow referred to the point of impact. A similar analysis would show how we manipulate the hammer for the next blow.

This causal picture, representing the policy, is highly dependent upon external circumstances. The size of the nail, the weight and balance of the hammer, whether the wood is hard or soft: all of these details require a different policy to accomplish the purpose. The purpose is constant; the analysis by which it is translated into the policy is highly variable.

But this analysis plays no part in what we actually do. It appeals only because we are so deeply immersed in the causal myth that we ascribe to it a reality going beyond its function as our explanatory description. If we are asked to explain in causal terms how we hammer a nail, we are quite unable to do so, because we do not translate our purpose into the policy in order to fulfil it. The same is true for other kinds of skill, including those which have a strong intellectual content such as engineering design or medical diagnosis.

For the programmable digital computer the case is different. Its actions do not arise directly from its fundamental purpose—a purpose of transistors and electrical circuits. We have intervened to incorporate a human purpose, subordinate to this fundamental purpose. In an 'expert system', for example, this has to be done by a process similar to that described in Section 10 of Chapter 5. The 'expert' is questioned by a 'knowledge engineer', who asks for a causal description of his behaviour; that is, for the policy by which his purpose is translated into action. The need for this step is indeed presented as an advantage, as a step towards the refinement of our knowledge conceived as expressible only in causal terms:

> Yet if artificial intelligence research had done nothing else, it had shown how empty most theories of intelligent behaviour were (likewise theories of creativity, originality, autonomy and consciousness). When you wanted to make a computer behave intelligently, you had to have a very precise idea of intelligent behaviour in order to specify it to the computer in detail. In neither psychology nor philosophy did such precise models of intelligence exist.[6]

In terms of the causal myth we do not understand intelligence or creativity or skill until they have been expressed in causal terms. That expert systems force us to do this therefore appears as a contribution to understanding:

> The heuristic [sc. problem-solving] knowledge is hardest to get at because experts—or anyone else—rarely have the self-awareness to recognise what it is. So it must be mined out, one jewel at a time. The miners are called *knowledge engineers* ... the expert himself doesn't always know exactly what it is he knows about his domain ...[7]

> What the masters [of their craft] really know is not written in the text-books of the masters. ... But we have learned also that this private knowlege can be uncovered by the careful, painstaking analysis of a second party, or sometimes by the expert himself ...[8]

Yet when we accept the purposive myth, as we do in this chapter, the work of the knowledge engineer appears as a *post hoc* rationalization of something that was, in most cases, done spontaneously without conscious analysis. Whereas the causal myth would insist that an analysis was nevertheless carried out, below the conscious level, the purposive myth sees the action following directly from the purpose, without prior translation into the policy. Both views are explanations constructed for human purposes of understanding and control. They are equivalent for scientific and technical purposes, but nevertheless they lead to quite different explanations and different courses of research and development and action.

Accepting the purposive myth we shall see, for example, great difficulties in what a knowledge engineer sets out to do. An expert has a purpose, and without analysis fulfils the purpose by his actions. These actions constitute a policy, but they will be different in different circumstances. And if circumstances arise which were not foreseen when the 'expert program' was developed, the policy generated by the computer may be quite different from the actions which an expert would think to be appropriate.

Because we are so deeply imbued with the spirit of the causal myth, it may be difficult to grasp the distinction which is being drawn between the behaviour of computers and of people. It can perhaps be made sharper by considering the behaviour of faulty systems. When a computer fails, either through a hardware fault or a software error, its behaviour no longer expresses the human purpose which was induced in it, but a different purpose arising from its faulty condition. Failures in the human organism can equally result in aberrant behaviour expressing the purpose of a faulty organism.

The two kinds of behaviour, resulting from human or computer failure, will be quite different. We could perhaps cause the healthy, working computer to simulate human failures, but we cannot make the failed computer behave in the same way as the failed human organism. The computer is like an actor who has learned a part. His 'knowledge base' is the script, together with his experience of acting. On this foundation he creates a simulacrum (but only a simulacrum) of a character whose words and actions are not scripted, but arise spontaneously. A sane actor can represent a madman: a mad actor represents only himself, and is spontaneously mad in his own way.

Expert systems are one special kind of computing system, but what has been said of them can be transferred without much change to systems designed on alternative lines. Algorithmic programs following a set course of calculation, clearly require a causal description of what is to be done. We could program into a computer some variational principle such as Hamilton's, and cause the computer to base its operation on this: but to do so it would first have to generate the policy. Computers can be made to 'learn from experience', but the same remark applies. 'Experience' would have to be incorporated in causal relations before it could be used.

In all cases, it is only necessary to remark that a knowledge engineer would never have the same problem in 'mining' the knowledge of a computer as was described above in the capture of knowledge from the expert. Somewhere within the software of the computer would be found an explicit causal description of any task which the computer executed.

These comments apply to programmable computers. There are some computer-like digital systems which do not use a program: for example certain devices employed for pattern recognition. The way they behave has been incorporated in them by a process similar to the one described in Section 10 of Chapter 5, but their behaviour is more obviously machine-like (in the ordinary sense) than the computer's. No strong claims to human abilities are usually made for these devices, which set out to imitate, more or less closely, a causal description of the functioning of the brain, rather than the mind.

To summarize what has been said, when we accept the causal myth, 'man is a machine' in the sense that every feature of his behaviour can be described in causal terms. Since anything which can be described in this way can in principle be done by a computer, it becomes hard, if not impossible, to say that any behaviour shown by man cannot also be shown by a computer.

When we adopt the purposive myth, 'man is a machine' in the sense that every aspect of his behaviour arises from an ultimate purpose. Every aspect of the behaviour of a computer equally arises from an ultimate purpose, but this differs from a human purpose. One arises from the physical properties of a vast collection of elementary particles composing a body of flesh and blood. The other arises from a vast collection of elementary particles composing transistors, electric circuits and the like. Human ingenuity can incorporate some parts of a human purpose in a computer. But it exists there as subordinate to an ultimate purpose very different from that of the human being. At some point, therefore, human and computer behaviour will always diverge. The difference of view between the causal and purposive myths stems ultimately from the fact that the former builds up the world from simple, self-sufficient entities, while the second begins from the opposite extreme, from the universal purpose.

The tendency of this discussion is not to suggest that computers should appear less practically useful when we accept the purposive myth. They give us the opportunity to free ourselves from much routine intellectual work, as an earlier stage of mechanization freed us from much physical labour, and their usefulness can be greatly increased by the new techniques which have arisen from AI. When they do not interact with people, the limited human purpose which can be incorporated in them can be adequate, and need offer no problem.

But where computers and human beings must interact, the purposive myth suggests that computers must be subordinate to men and women. Human purposes can never be incorporated in computers in the way that they exist in people. As the surrounding environment changes, the policy derived from the purpose will also change, and at some point the policy generated by the computer will differ from that which would be generated by the human organism. Unless the computer is subordinate to human judgement, those who interact with it will find themselves helplessly watching as it pursues the consequences of its error.

The role of the computer must therefore be to assist and support the human skill with which we achieve our purposes, for example by suggesting courses of action for human assessment, and by warning of the danger of proposed actions. It must do this in a way that allows human skill to evolve and develop, rather than seeking to replace it. More will be said on this subject in Chapter 9.

7. *Formulating human policies*

The discussion which has been given is somewhat complicated by a fact that we have ignored, namely that we do ourselves, on occasion, translate a purpose into its policy in order to implement it or describe it. Indeed, we regard our ability to do this as one of the most important attributes raising us above the other animals. The chief circumstances in which we make this translation seem to be the following.

i. When the fulfilment of a purpose requires the cooperation of a number of people. If, for example, we are searching for something which we have dropped in the street we might say to a companion, 'I'll take the left-hand side, you take the right.' The resulting action subdivides the purpose of searching the street into two sub-purposes: of searching the right-hand side and searching the left-hand side. Taylorism (Chapter 8) attempts to take this kind of subdivision to such a point that individual tasks are reduced to a small sequence of simple operations, repeated continuously and exactly.

ii. After the event, we may need to describe and justify actions, which we took in order to fulfil a purpose. After a motoring accident, for example, we may reconstruct our actions for insurance purposes, saying how we saw a car emerge from a side-road, and braked hard in a straight line, but then had to relax braking in order to eliminate an incipient skid as we steered to the right. When they occurred, our actions were not analysed in this way—

they took place before we were consciously aware of the need for them—and it may demand some effort after the event to reconstruct them. The process by which an 'expert' explains his actions to a computer scientist constructing an 'expert program' is similar. So also is the process by which a skilled driver, unused to teaching a beginner, reconstructs a sequence of actions for the beginner to follow.

iii. When we are learning an unfamiliar task, we may begin by breaking it down into successive actions. When we are learning to drive a car and wish to turn a corner, we may follow a list of actions which we have been given: release throttle, depress brake, depress clutch, move gear lever, release clutch, turn wheel, etc., etc., in a conscious sequence. But after a time we no longer perform the actions in this way, but as a combined motion which fulfils our purpose.

If we believed the causal myth, we should say that the sequence of individual actions is stored somewhere in the brain, but is no longer conscious. As we are adopting the purposive myth, we say that the purpose is now fulfilled without being first translated into the policy. There will always be a description of our actions in terms of the policy, and we can with some effort generate this, as in (ii). The policy will vary with such circumstances as whether the road is dry or wet or icy. But our reaction to these circumstances will not occur by a logical analysis. We shall be aware, for example, of the grip of the tyres by the force needed to turn the steering wheel, and the incipient sliding at front or back. But these statements represent an explicit analysis which we do not make at the time: we 'feel' what is happening between tyres and road, in the same way as we 'felt' what happened at the head of the hammer when it hit the nail. The car has become an extension of ourselves, like the hammer, and its purpose has been, to some degree, incorporated in our own.

The purposive description of human behaviour illuminates some aspects which cannot be easily understood in a causal description. One of these aspects is the common experience of problem-solving, in such activities as design or the generation of new mathematical theorems. It is generally recognized that these activities, though purposive and highly structured, cannot be reduced to a formal procedure.

A common description of the way that problems come to be solved is that one must worry over them for a long period, usually without much success, but then it is best to put them aside. Some time later, without any further thought, a solution will often present itself. It needs to be checked, because sometimes it is wrong, but often it is correct.

The decisive step is to internalize the problem to the point where one is committed, at a deep level, to solving it. This purpose of obtaining a solution is then fulfilled by the human organism, though we have no conscious knowledge of the way this is done. Nor can we propose any logical procedure by which we can be certain to generate the result; though it is easy to formalize the process of verification and proof once a tentative result is available.

If we accepted the causal myth, we should say that at a subconscious level, some logical procedure was carried out within the brain to produce the required answer. Accepting the purposive myth, we say that our organism fulfils the purpose which we have formed, and which is a subordinate purpose of the whole human organism. Some causal description of what went on can be obtained, in principle, by evaluating the policy obtainable from our purpose, but this description is a description and no more. It is not the mechanism by which the answer was obtained.

8. *The wider implications*

In this chapter we have given a brief sketch of the world as it would appear to us if we accepted the purposive myth. This explains all phenomena as the consequences of a purpose. The purpose is pursued, generally in the face of random disturbances, and is accomplished to the greatest degree possible subject to these disturbances. The actions which are needed to fulfil the purpose constitute the policy, and are in the form of causal relations: given particular circumstances at a particular time, the required action is prescribed by the policy. An alternative description of the world takes these causal relations as basic, and ignores or rejects the purpose.

Both descriptions are valid to the same extent. They are equivalent in the sense that evidence which supports one supports the other and evidence which contradicts one contradicts the other (the demonstration of this fact being incomplete in certain respects, but in principle amenable to completion). From a scientific point of view it should be a matter of indifference which of the two we choose to use, or at most a matter of convenience. Yet for our relation with the world, and our behaviour in it, the two views clearly have different implications.

On the causal view, nature has no purpose. Human beings appear to have purpose, by direct introspective knowledge, and living things by extension may also have purpose. But this purpose exists only in a shadowy form. Upon analysis, it dissolves into causal behaviour, and reflects no deeper underlying purpose. At most, it is the result of natural selection acting upon chemically determined and chance-mutated matter, and is the natural characteristic of whatever survives: it is the badge of the survivor.

In such a world, man is a stranger, but a liberated stranger. He has no duty, because duty implies an ultimate purpose. He has no responsibility to the future, because the future is as meaningless a concatenation of events as the present. Meanwhile his own purpose exists to himself; and the world, lacking any purpose, is subject to his will to the extent that he can impose it. The world exists to be owned and used and exploited, and in the pride of imposing his will he can forget the mental anguish of inhabiting such a world.

This describes a sickness, deeply felt perhaps only by a few. But if we regard the way in which modern societies act, we shall find that it is in full accordance with the view described. We are exploiting the present without care for the future, and exploiting the future itself to the extent of our power.

This is the view of the world corresponding to one myth. The other myth describes the world by means of the purpose which it follows. This is not something known a priori, or imposed theologically. It is the result of our attempt to explain the world, and the form which it takes discloses no hidden meaning. We have to accept it as a human construct, resembling in that respect the causal myth, and equivalent to it.

But though it is equivalent in the strict sense explained in Chapter 4, the purposive myth engenders a very different climate of belief about the world from the causal myth. Human purpose is not something alien, not just an appearance created by the evolutionary pressure in those organisms which succeed in surviving. It is part of an inherent purpose in the world, shared with animate and inanimate nature. We are at home in the world, but do not own it: we belong to it, but it does not belong to us.

Consequently we are no longer free, as the causal myth suggests, to do as we wish with the world, up to the limit of our power. As part of an organic whole we have a responsibility for the whole. We face again the ancient quest: 'what it was good for the sons of men that they should do under the heaven all the days of their life', a quest to which no answer is supplied.

The two views of the world will govern our behaviour in it, and this is the topic of the succeeding chapters. What we have been mainly concerned to establish, up to this point, is that both views are man-made. Both arise out of our investigation of the world and our attempt to explain it. The choice of one or the other is made at the very beginning of this attempt by our decision about what kind of explanation we shall accept. If we agree with Monod[9] that 'For science the only *a priori* is the postulate of objectivity [sc. causality]' then this a priori will lead inescapably to a causal description of the world, which we can support by showing that it explains all observed phenomena.

If on the other hand we accept that causal relations can represent the policy which will accomplish a purpose, then we can end with a description of the world in terms of purpose. This we can again support by showing that it agrees with all our observations. In each case, the condition which we impose initially on the kind of explanation we shall accept, appears again at the end as a consequence arising from the nature of the world.

It will now be clear why the causal description and the purposive description have been called 'myths'. Like more primitive myths, they strongly determine our view of the world and the way in which we should behave. What is new, or at least far more elaborately worked out, is the factual structure and theoretical superstructure of science by which the myths are supported.

References

1. Maurice A. Biot (1970). *Variational principles in heat transfer*, Clarendon Press.
2. See for example B. W. Hyndman (1987). Cardiovascular system: baroreflexes model. In Madan G. Singh (editor), *Systems and control encyclopedia*, vol. 1, pp. 536–42, Pergamon Press.
3. Jacques Monod (1971). *Chance and necessity*, Collins Fount.
4. Hubert L. Dreyfus (1972). *What computers can't do*, Harper and Row.
5. Michael Polanyi (1958). *Personal knowledge*, p. 55, Routledge and Kegan Paul.
6. Edward A. Feigenbaum and Pamela McCorduck (1984). *The fifth generation*, p. 52, Pan Books.
7. Reference 6, pp. 104, 114.
8. Edward A. Feigenbaum (1979). Themes and case studies of knowledge engineering. In Donald Michie (editor), *Expert systems in the micro-electronic age*, p. 8, Edinburgh University Press.
9. Jacques Monod (1970). *Chance and necessity* (translation 1971), p. 98, Collins.

7 The causal myth

How it is now

1. *Repercussions of the causal myth*

The aim of the previous chapters has not been to persuade the reader to any belief in the purposive myth. It has been instead to dissuade from a belief in the causal myth, which has a fierce and all-embracing grip upon every modern industrial and scientifically based society. So powerful is this grip that we can only with the greatest difficulty see it for what it is—one among many ways in which we can explain and relate to the world.

We are led by every influence to regard the causal myth as the ultimate and only truth about the world. We meet it first as young children, and become more and more familiar with it through our schooling and our daily experience. It is not only, nor even mainly, through formal education in science that it is impressed upon us—indeed a scientific training may lead to some beginnings of scepticism. It is rather an all-pervading influence borne in from every side, and supported by every observed fact to which we can assign an explanation. For our explanations are causal—whether in physics or chemistry or biology or social science—and to the extent that they are verified by facts they suggest that nature itself is inescapably causal. Medieval society in Europe had much the same relation to Christian theology.

The only way to appreciate the hold which the causal myth has upon us is to present an alternative myth. This is the service performed by the myth of purpose. It is so constructed that the same consequences can be deduced from it as from the causal myth; which indeed simply embodies the policy which will achieve the purpose. To the extent that the purposive myth can be carried on and elaborated in this spirit, no conflict will ever arise between it and the causal myth. They will be scientifically equivalent, in the sense that all evidence which supports one will support the other, and all evidence which contradicts one will contradict the other.

2. *Our relation with nature*

Yet despite their scientific equivalence, the two myths will induce in us quite different outlooks and quite different kinds of behaviour. The demonstration of this difference will be made most thoroughly in Chapter 8, in relation to the working conditions created by our technology: it is in that area that the whole investigation originated. But to show how widely the effect of the causal myth extends, a number of other examples will be given here more briefly.

Let us start with a scene from a fertile valley in California. The road ran straight across flat open country, raised a little above the general level, and with sloping banks to either side. In every direction, the fields stretched away without a break for mile upon mile. There were no dwellings, and no workers to be seen in the fields; no-one walking upon the road, and only the occasional car. At other seasons there would doubtless be farm machinery and labourers upon the land.

The fields were green with crops: lettuce, cabbage and the like in close endless rows. Here and there were irrigation hoses sending up bright arching jets of water that sparkled under a brilliant sun in a clear blue sky. It was a scene of prosperity and order and plenty.

But there were no trees, no hedgerows, no plants of any kind except the regimented crops. The earth on the banks of the road was scraped bare and clean. Weeds no doubt use water. There was no sound of insects. They no doubt were controlled by chemical sprays. And with nothing to feed upon, there were no birds. It was in fact an empty landscape, dead except only for the crops which grew in their carefully tended lines.

Monod[1] describes us as living in 'a world of icy solitude', deaf to our music and as indifferent to our hopes as to our sufferings or our crime, and this Californian valley would be an apt setting for such a life. Izaak Walton[2] had no doubt a romantic view of nature, very different from the view of a peasant oppressed by endless labour. But when he told of his 'sweet content', as he sat quietly, watching 'here a boy gathering lilies and lady-smocks, and there a girl cropping culverkeys and cowslips', he was describing something that the peasant would have recognized, from his own and his children's childhood. But no children will ever play or gather flowers in that bright, sad valley in California.

The contrast is too bleak, throwing a dark shadow over our future. But if we wished to object to it, on what grounds could we do so? We should be told, and rightly, that scientific farming on the Californian plan is what has saved a great part of the world from starvation. That the mechanization which accompanies it is what has alleviated much (not yet all) of the desperate, back-breaking toil that once was demanded by the land. That we know no other way of providing, for the majority, a life of relative prosperity and plenty, with moderate labour and generous leisure.

To object to farming in the antiseptic, clinical style which makes man an outcast and a stranger in the land is equated with a wish to return to past conditions: to a state where, even without the added miseries of war that Hobbes[3] supposes, there were for most 'no Arts; no Letters; no Society; ... And the life of man, solitary, poore, nasty, brutish, and short.' Within the framework of our modern thought, it is very hard to maintain an objection against this criticism. If the means we use are the only ones that will achieve them, then to reject the means is to reject the benefits.

Yet when we analyse the criticism, we can see that it is based on a preconception of causality. We wish for a more productive agriculture than existed in earlier times. We must therefore study nature in the scientific spirit which regards it as a machine, subject to causal laws, and available to be manipulated in any fashion we choose, to fulfil our own purposes.

The last sentence does indeed express an inescapable conclusion if we believe that 'The cornerstone of the scientific method is ... the systematic denial that "true" knowledge can be reached by interpreting phenomena in terms ... of purpose.'[4] This gives us the causal myth. But if we believed the purposive myth, we could approach nature in a different spirit: one in which we respected the purpose expressed in other living things, and sought to co-operate with it to achieve our own purposes, rather than to coerce it by main force. And if we see the causal interpretation of nature and the purposive interpretation as two equivalent myths, distinguishable from the theory to which they both belong, we are released from the compulsion to approach nature in a spirit of violence.

3. *The Lushai Hills*

Four hundred years of belief in the causal myth have given us a technology, including our agricultural technology, which embodies the values incorporated in the myth. We have great difficulty in imagining a different technology which could give equivalent benefits, or in believing that such a technology is possible. What is, seems to be all that could ever be. Only by showing an example of a better technology, or at least the basis on which it could be developed, can we hope to overcome the disbelief in its existence.

This will be attempted, for one particular technology, in Chapter 9. Meanwhile, we can at least try to show how what has been done in the past four hundred years has come to seem the only possible course which could have been followed. To this end, consider a metaphor which has been used before.[5]

The Lushai Hill Tracts, in the early 1940s, were almost untouched by any human activity. Long ranges of hills ran north to south, rising gently at their western boundary to a few hundred feet, then falling again almost to sea level in a river valley. Range upon range succeeded to the eastward, each ridge climbing higher, but falling again into a deep valley, until at Blue Mountain the highest peak reached 7000 feet.

Scattered settlements were confined to the crests, partly to escape the malaria of the lowlands, and partly from a tradition of defence; the Lushais had only recently given up their immemorial practice of head-hunting. Their villages contained perhaps thirty or forty inhabitants, living in bamboo huts, and growing hill rice upon land fertilized by burning off the brushwood. After a few years, the poor soil would be exhausted, and the villagers would move on, taking with them their chickens, their few pigs and goats, and their dogs.

For the most part, the hills were covered in jungle. Almost everywhere were tall trees providing a canopy, with dense undergrowth, with here and there occasional patches of close-growing, unshaded bamboo. The war mostly passed by the area, with only forays by small parties, or feuding between tribes attached to either side.

It was a region of great natural beauty. Following one of the major rivers by canoe, in the brilliant sunshine of noonday under a cloudless sky, there would be a complete silence, broken only by the distant braking call of monkeys. On either side the dense foliage would rise up steeply, covering the hills to their summit.

Travelling north to south along the valleys was easy, but not so the passage west to east. There were occasional tracks or game trails, but often the only practicable routes were along the streams which cascaded down the hillsides, forming low tunnels under the overhanging foliage.

Climbing upwards in this way, one would reach a fork where two streams joined, and a choice had to be made. No reliable information could be obtained from the map, and no general overview was possible to guide the choice, which must be based only on what could be seen within a few yards, or on any general predisposition to go towards the right or the left.

For a while, it was possible to reverse a decision if it proved unfortunate; either by going back to the junction or perhaps by cutting across from one stream to the other. But very soon this became impracticable, because of the great effort needed. One was committed to the chosen stream, for better or worse.

Having climbed high up the side of the valley, one would pause and camp for the night. Looking back to the west, one would see range after range of hills, falling away to the plain. The red of the setting sun would cast a glamorous light over the country from which one had come, covering the hills with a purple haze and disguising the heat and the malaria and the leeches from which one had escaped.

Then it was possible to feel a sense of achievement: to have climbed so high and to be able to look back over the lower country out of which one had come. And it was easy to believe that all the choices which had been made along the way were justified by the outcome, and were the only right choices to be made.

This self-congratulation might of course have been quite unwarranted. Some other route might have led to still higher ground, and done so more easily. But if so, the knowledge was hidden, and the complacency uncontradicted.

This is an image of the way our technology has developed. We have climbed by our efforts out of a past in which the relative ineffectiveness of their labour condemned the great majority to unremitting toil and poverty; and looking back there is even a glamour upon this past which conceals some part of its harshness. In our progress, we have continually had choices before us. We have been ignorant of their ultimate consequences, and have taken our decisions by the light of the causal myth.

The advantages we have gained cannot be given up lightly, nor can we, if we wished, go back again and start from where we began. We arrived where we are by the route that we chose, and knowing no other are persuaded to believe that it is the only one we could have followed with success. Yet Monod's 'abyss of darkness'[6] which opens before us must cause us to ask whether this appearance of inevitability arises only because we chose one particular guide in making our decisions; whether we might not have climbed as high by some other route, but arrived in a more welcoming countryside.

4. *Man and the animals*

The unease and disquiet which are lightly brushed by the dehumanization of the countryside are touched more sharply in a second image. It shows a laboratory, in a pharmaceutical company, devoted to the testing of drugs. There is none of the chemical equipment one might expect, of benches and bottles and flasks. Instead there is a room about fifteen feet square, with two walls lined by tall racks of electronic equipment.

From a loudspeaker comes the muted sound of a heart-beat, keeping time with the upward leap of the trace which slowly traverses a cathode ray tube. Recorders gently unroll wide charts of paper under recording pens. Wires lead from the racks to a wooden table, in the centre of the room, where they are connected to a dog lying inertly upon it.

The dog is anaesthetized, and lies upon its side. Its chest has been opened, and through the gaping hole air is drawn in and expelled as it breathes. One is surprised at first by the absence of all the sterile precautions of an operating theatre—the masks and white coats, bare walls and metal furniture. But surprise is dispelled by the realization that the dog will not regain consciousness: it will be painlessly killed when the experiment ends.

The aim of the experiment is to test a new drug for its possible effects upon the heart. It offers the hope of alleviating human disease, but before it can be used for this purpose its safety must be checked. This is being done by gradually increasing the dose to which the dog is subjected, in order to find the point at which an adverse effect upon its heart is detected.

To the unaccustomed, the sight of such an experiment is likely to be distressing. No objection can be made on the ground of cruelty. Anaesthetizing the animal requires no more than the slight prick with a needle to which human candidates for surgery are accustomed. There would be no anticipation of injury or death, and no consciousness after the first moment.

Yet when this is acknowledged, the disquiet remains. A dog is a social animal, and we are social animals of a different kind. We share certain types of behaviour which the social role imposes, and beyond the profound differences can directly interpret the expression of them: of submission and ingratiation, joy and fear, courage and affection and willing co-operation. And over many millenia, mankind has achieved the slow transfer of much of the dog's social behaviour from its own species to human kind.

Then when a dog is treated as an object for laboratory experiment, there must be sense of betrayal. We may feel that direct affection of man for dog beyond a certain limited point is unseemly, but there is a mutual recognition and acknowledgement which ought not to be denied.

That it is denied in laboratory experiments is clear, and the origin of the denial in the causal explanation of nature is equally clear. Animals are machines, if they are causally explained, as was pointed out by Descartes. After remarking on the success which had been achieved in the construction of automata, even with simple means, and contrasting them with the much more complicated and highly organized bodies of animals, he continues,

> if there were such machines [sc. automata], having the organs and the outward appearance of a monkey, or of some other animal not possessing reason, we should have no means of telling that they were not of entirely the same nature as these animals.[7]

And to experiment upon the machine which is a dog, for the benefit of humanity, need raise no objection in the most tender conscience.

We meet here the same argument that we considered in relation to agriculture: the development of drugs through laboratory testing upon animals has led to an immense increase in knowledge, and in our power to alleviate human suffering. We know of no other way of obtaining this knowledge and power, and if we reject the means by which it was obtained we are at the same time rejecting the benefits it brings.

To this we can answer as before. The route which we chose in order to acquire our knowledge was determined by our choice of a causal explanation of nature, and by our denial that any alternative explanation was possible. If an alternative explanation can be given—an explanation in terms of purpose—then we shall expect the pursuit of knowledge by its means to progress by a different route: one which perhaps will not result in any less material benefit but will avoid the moral unease that we now incur.

5. *Production systems*

As a third example, consider a small plant for producing electric light bulbs having a metallized reflector. Production was 800 in the hour, using glass envelopes brought in from another part of the plant in two pieces. Working conditions were noisy and somewhat bleak, but the environment was not otherwise oppressive.

Manufacture of the lamp entailed a number of stages: metallizing the reflector, sealing in the filament, fusing the two parts of the envelope together, evacuating and filling with gas. Lamps travelled around the periphery of the production machine on a chain conveyer, and most of the process was automated.

For example, where the two parts of the envelope were to be sealed together, the glass body was transferred to a rotating table. This indexed through a number of positions, where the body was first pre-heated, then met with the front portion, and was fused to it before being returned to the chain conveyer.

At certain points, however, the designer of the machine had not been able to achieve complete automation, and here there were workers, mostly women, mostly middle-aged. One picked up the larger part of the envelope at the beginning of the process and inspected it for flaws or cracks, before replacing it on the conveyer—a task which is probably still beyond the abilities of computer vision and which, when it becomes possible, will at first be expensive.

Eight hundred bulbs an hour equates to one every four and a half seconds. The woman stood at the machine, isolated from other workers, unable to talk to them because of the noise, and with a part of her attention continually devoted to her task. It is highly unlikely that the designer of the machine had ever imagined to himself the life he was creating for the worker. Most probably, in his mind, the worker was equated to a machine component in a way that was noted as early as 1833:[8] 'The labourer is indeed become a subsidiary to this [sc. steam] power . . . he is condemned . . . to become himself as much a part of its mechanism as its cranks and cogwheels.'

At a second point the conveyer was at bench height and another, rather younger, woman was seated and leaning over it. Her task was to take, for each bulb, a short length of aluminium wire by one end in tweezers. Then she had to insert it within a coil, placed in the bulb, which would afterwards vaporize it in order to metallize the glass of the reflector. It was only just possible to carry out this task in the time available, so there was some sense of strain. Occasionally the wire was dropped, or incorrectly inserted in the bulb, which would have to be removed from the line by a male supervisor.

In these tasks, the statement that 'man is a machine' takes on a meaning which its originators can hardly have envisaged. The statement was made in the context of science, but life is not so neatly compartmentalized as our intellectual activities. In particular, as will be shown in Chapter 8, technology borrows its ideas and its values very directly from science.

The distaste aroused by the sight of this plant springs partly from a sense of degradation and waste: that human abilities should be reduced to so low an estimate. Partly also it springs from the imagination of what such work, day by day, year after year, must do to human spontaneity and liveliness and interest. The literature of social science[9] contains many a lamentation on this theme, but few have expressed themselves as directly and brutally as Adam Smith:[10]

> In the progress of the division of labour, the employment of the far greater part of those who live by labour, that is, of the great body of the people, comes to be confined to a few very simple operations; frequently to one or two. But the understandings of the greater part of men are necessarily formed by their ordinary employments. The man whose life is spent in performing a few simple operations . . . has no occasion to exert his understanding . . . He naturally loses, therefore, the habit of such exertion, and generally becomes as stupid and ignorant as it is possible for a human creature to become.

A more inspiriting condemnation comes from the recollections of Tom Bell:[11]

> I remember Arthur MacManus describing a job he was on, pointing needles. Every morning there were millions of these needles on the table. As fast as he reduced the mountain of needles, a fresh load was dumped. Day in, day out, it never grew less. One morning he came in and found the table empty. He couldn't understand it. He began telling everyone excitedly that there were no needles on the table. It suddenly flashed on him how absurdly stupid it was to be spending his life like this. Without taking his jacket off, he turned on his heel and went out, to go for a ramble over the hills to Balloch.

6. *Recapitulation of the argument*

Before proceeding it will be useful to recapitulate the argument which is being developed in this chapter. This is, that our long tradition in science accepts no explanations except those in terms of cause and effect. The result is that we regard everything outside ourselves as a machine, and a machine without purpose.

In principle, we also regard ourselves as machines, but the strain of doing so proves in practice to be too great. So we accept our own purpose without reservation; we have also a social life within which we set aside the need to reject purpose. Yet in all serious matters of science and technology and commerce we act as though our own purpose were unique.

The examples used above in illustration are taken from personal observation. In agriculture we treat the natural world more and more as though it were an inert, purposeless machine, governed by causal laws which we can elucidate, and which we use to attain our own purposes. We take the same view of animals, treating them also as machines. We design our production systems so that human activity is made to resemble as closely as possible the activity of a machine. If these statements seem exaggerated and extreme, a consideration of tendencies in the last few decades will confirm their accuracy. Some fringe reactions apart, all movement has been in the direction suggested. Where the statements do not apply, we usually find a survival of previous conditions, not yet amenable to the mechanizing tendency.

In the naive (but not thereby less perceptive) observer, the more extreme examples of this tendency will raise misgivings, some of the bases for which have been indicated above. In agriculture it produces a landscape which repels all human feelings of closeness, of belonging, of delight: a dreamlike landscape as of some other planet. It sets a gulf between ourselves and the animals, severing a bond which we feel though we cannot easily articulate. Between those who design production systems and those who work in them, it sets up a destructive antagonism, in which a human purpose of production is achieved by denying purpose in humanity.

These feelings are probably becoming more widespread and stronger, but to be effective, rather than merely destructive, they need an intellectual basis from which something more satisfactory can be evolved. The difficulty of providing this basis is considered in Section 8.

It is also necessary to reply to the objection that all the progress we have made in alleviating poverty and hunger and disease has been achieved by exactly that scientific outlook which is in question. If we reject it, we reject at the same time the advantages which it brings. A partial reply has been given in Section 3: that the success of one path to the acquisition of knowledge does not prove that there cannot be other, and perhaps better, paths. As a logical proposition this is sound enough, but it evokes the response: 'Show me'. An attempt to show the basis on which this could be done, for the particular case of industrial production, is made in Chapter 9.

7. *Science and medicine*

Before considering the matter more generally, it is instructive to look at a further example, which is provided by the relation between science and medicine. There is in medicine a long-standing and admirable commitment to an ethical standard of behaviour. There is also a long tradition of scientific study from which great benefits have been obtained. Yet the relation between ethics and science, which here means the causal myth in science, is an uneasy one.

To the physician acting within the ethical tradition, the patient is a fellow human being to whom he can give help and comfort. Nothing is to be done to the patient that is not aimed solely at his own benefit, unless it is done with his own free and informed consent. The medical scientist has a different but equally admirable aim: by increasing knowledge, to permit the more accurate diagnosis and more effective treatment of medical problems.

One might expect that these two aims could coexist without conflict, but there is a tension between them. Science, as always, sets up a distinction between the intelligent and purposive investigator, and the inert, causally determined object of study. Thus Pappworth[12] remarks on 'the significant fact that in most medical reports of patients having been submitted to experimentation, the patients themselves are collectively described as "the material".'

With this outlook, the scientist departs from that attitude to the patient which must underlie and support the ethical requirement. The contrast can become explicit:[13]

> The desire to alleviate suffering is of small value in research—such a person should be advised to work for a charity. . . Research wants egotists, damned egotists, who seek their own pleasure and satisfaction, but find it in solving the puzzles of nature.

It is therefore not surprising that conflicts can arise between the scientific and ethical compulsions, a conflict in which the initiative is ordinarily with science. Some examples will illustrate this situation.

(*a*) A research project[14] in the late 1970s proposed to improve the management of childbirth in the following way. A labour-inducing drug would be continuously injected into the mother while the heart-rate of mother and baby were monitored. If either showed signs of distress, the rate of injecting the drug would be reduced. This would be done by an automatic control system, having the aim of keeping the rate of injection at the highest level which just avoided distress. In this way, the delivery would be made as quick as it could possibly be with safety.

If it were successfully developed, and made available, a woman might well choose this method for its safety and speed, and there would then be no conflict with the ethical requirement. Nevertheless, the equation of the mother to a machine is clear. In a real sense, mother and control system become one machine, and it is this machine which 'gives birth'. Without questioning the aims of the research, one can have strong reservations about the means by which it was proposed to achieve them.

(*b*) Experiments in 1915 were reported by a doctor who

made an unsuccessful attempt to discover a cure for pellagra. To do this he produced the disease, which is characterized by diarrhoea, dementia and dermatitis, in twelve white Mississipi convicts who became seriously ill as a consequence. Before the experiment was made formal agreements were drawn up with the convicts' lawyers agreeing to subsequent parole or release.[15]

Here consent was obtained, but it cannot be considered as freely given. It had been purchased by a reward which might have great value for the prisoner.

(*c*) In March 1962 a forty-year-old married man

presented himself . . . for operation on a hernia. The previous year, at the same hospital, he had been found to be a diabetic, and was put on insulin. The account of this patient contains the following statements:

> At the time of the investigation into the diabetes he was recognised as having a genital deformity about which he was so sensitive that he failed to keep an appointment at which it was hoped to investigate this aspect of his case.
>
> For this reason he was not questioned about his marital relationship at the time of admission for the hernia operation. Neither he nor his wife volunteered any further information about the matter, and it was not thought to be in his interests to worry him further. It was decided to examine his genitalia very closely under anasthesia at the time of the hernia operation.[16]

Some female characteristics were found, the abdomen was opened, and a cystoscope was passed into the bladder to obtain further information. The anaesthetized patient was here treated as an inert object of study. No consent was obtained, and from the account it would clearly not have been given if asked for.

(*d*) The experiments by doctors on inmates of Dachau and other concentration camps are notorious and repulsive. One experiment was intended to discover the best way to resuscitate pilots who had been chilled in the cold waters of the North Sea.

The subject [an inmate of Dachau] was placed in ice-cold water and kept there until he became unconscious. Blood was taken from his neck and tested each time his body temperature dropped one degree . . . The lowest temperature reached was 19° Centigrade, but most men died at 25° or 26°. When the men were removed from the icy water, attempts were made to revive them . . . [by various methods][17]

Medical ethics here have been forgotten. Men are treated as objects, and objects of no concern. It will be thought objectionable and unfair to attribute the doctors' behaviour to a scientific attitude, since Nazi ideology had already robbed the prisoners of all human consideration. But one can at least say that science, in the mould of the causal myth, does not provide any intrinsic defence against being led in the direction which the doctors took.

The proceedings of the investigating tribunal make it clear that many scientists, not themselves involved in the experiments, listened without protest to accounts of what was done. The two German observers of the proceedings comment,

> Only the secret kinship between the practices of science and politics [sc. Nazi ideology] can explain why throughout this trial the names of high-ranking men of science were mentioned—men who perhaps themselves committed no culpable act, but who nevertheless took an objective interest in all the things that were to become the cruel destiny of defenseless men . . . This is the alchemy of the present age, the transmogrification of subject into object, of man into a thing . . .[18]

8. *Lack of a moral argument*

The reader will appreciate that the case which is being made against the causal myth in science is only in part the logical one which was made in earlier chapters. It also has a moral and ethical aspect which we have stressed in the present chapter. This being so, it would be helpful if assistance could be obtained from the prevailing world views of the present time. Unfortunately little help is available from this quarter.

In the Western tradition there are three main views to which we might appeal: Christianity, humanism, and Marxism. The first and last of these, though fragmented, are explicit and codified, while the second is more diffuse. Christianity in Europe has been in slow decline for two centuries, while Marxism has suffered a severe setback in the last forty years from its economic failures, and its continuing repressive character in those countries which have espoused it. Nevertheless, support from either, or from humanism, would be welcome.

An initial difficulty is that those who accept any one of these views, also in general accept the proposition against which we have argued in Section 3. They accept, that is, that the causal myth is the only basis on which it is possible to generate an understanding and control of nature, having the scope and power of our present science. In consequence, they believe that rejection of the causal myth involves the renunciation of those advantages which arise from scientific knowledge.

Few are prepared to make this renunciation, and the refusal to do so acts as a strong constraint on the conclusions which most will draw from their fundamental belief in Christianity or humanism or Marxism. Without the constraint, quite different conclusions might be drawn, but we cannot easily say what they would be. We therefore have to accept the conclusions as they are at present drawn, pointing out, where we can, the influence which has been exerted by the constraint.

With these preliminaries, we can consider what support, if any, can be drawn from the three world views for the objections, which we have raised in our examples, to some consequences of the causal myth.

i. None of the three appears to give any basis for objecting to the kind of agriculture described in Section 2. The Biblical account of creation can be quoted to support any use which we wish to make of nature:[19]

> ... and God said unto them, Be fruitful, and multiply, and replenish the earth, and subdue it: and have dominion over the fish of the sea, and over the fowl of the air, and over every living thing that moveth upon the earth. And God said, Behold, I have given you every herb bearing seed, which is upon the face of all the earth, and every tree, in which is the fruit of a tree yielding seed; to you it shall be for meat.

Orthodox Marxism is penetrated by a nineteenth century view of science (meaning by this the causal myth) as the conquest of nature in the service of man. Humanism is likely to stress the material benefits which have arisen from science, rather than any subjective unease which we feel. In all three cases, the belief that the causal myth is essential to scientific progress appears to be a determining factor.

ii. Similar arguments apply to the animal experiment described in Section 4. If this caused suffering to the animal there would probably be reactions varying from outright rejection, to a balancing of benefit (to humanity) against suffering (of the animal). But as we have noted, no suffering is involved.

iii. Support for objections to the lamp plant can be obtained from all three world views. The social science literature contains numerous criticisms of the kind of work described, based essentially upon a humanist concern for the damage which it does to those who engage in it. Marxism gives rise to criticisms which are still more strongly expressed,[20] the source of the evil being traced to the capitalist mode of production. A Christian concern was expressed in 1931 by Pius XI in Quadragesimo Anno:[21] 'For from the factory dead matter goes out improved, whereas men there are corrupted and degraded.'

Yet the support is less than wholehearted. The designer of the lamp plant reduced its workers to the status of programmed automata. In this he followed the precepts of F. W. Taylor's *Scientific Management*:[22]

> Under our system the workman is told minutely just what he is to do and how he is to do it; and any improvement which he makes upon the orders given to him is fatal to success.

Lenin's assessment of Taylorism in 1981 was double-edged:[23]

> The Taylor system ... is a combination of the refined brutality of bourgeois exploitation and a number of the greatest achievements in the field of analysing mechanical motions during work. We must organise in Russia the study and teaching of the Taylor system and systematically try it out and adapt it to our ends. At the same time . . . we must take into account the specific features of the transition period from capitalism to socialism, which . . . require the use of compulsion . . .

In fact, after a brief debate[24] in the early 1920s, Taylorism became the accepted doctrine in Russia and has never since been effectively questioned; and the lamp plant was operating in a socialist state.

In a similar way, principled objection on Christian grounds to the working conditions exemplified by the lamp plant has never been very strong or effective. Criticism on humanist grounds has been continuous, but rather plaintive than directed to action. From all sides, criticism has been muted by the belief that the objectionable features of much industrial work were the regrettable but necessary condition for its economic effectiveness.

iv. General support for the ethical code in medicine is no doubt forthcoming from all of the three views. Beyond this, they seem to have little to say regarding the examples (a), (b) and (c). Very strong condemnation of (d) can be assumed, but there it is least needed because condemnation will be nearly universal.

These comments are all directed to Christianity and humanism and Marxism as systems of thought. They are not intended to apply to individual believers in these systems, among whom will be found many of the strongest opponents of the things which have been criticized.

9. *Conclusions*

If the analysis in the preceding section is correct, no great assistance can be obtained, from any world view current in Western countries, in making a moral or ethical case against the causal myth and its consequences. The conclusion is not surprising: it is not to be expected that the predominant view, in any industrialized country, will be one which provides a damaging criticism of its scientific and technical foundations. There then remain three arguments which can be used:

i. That the empirical evidence which supports the causal myth also supports, and to the same extent, the purposive myth. This was the aim of earlier chapters. The effect of the argument is to show that the causal myth is a human construct. When we find that it is confirmed by every observation that we make of the world, we are rediscovering what we put there to be found.

ii. That a technology based on the causal myth and incorporating its values is unsound, and less effective than alternatives. We can, for example, argue that the agricultural practice described in Section 2 provides ideal conditions for the proliferation of pests and disease. We may control these in the short term by chemical means, but mutant strains will continually arise to penetrate our defences. We are, in effect, pitting our own sole purpose against the purpose of all those organisms which can exploit the conditions we have created; and in any contest of this kind we are likely, in the long term, to be defeated.

Such an argument has to be made in detail, and case by case. It can be made[25] in relation to the Tayloristic development of production systems using computers, but as will be stressed in Chapter 9 there is a difficulty. This is, that the argument cannot be made within the causal framework without implicitly accepting the values which underlie it, and these by a logical development lead back again to Taylorism.

iii. That the naive and untutored reaction to many consequences of the causal myth is one of unease, disquiet, disgust, or desolation. When it is believed that there is no effective alternative, these feelings are held in check. If it can be shown that there is an alternative, then the feelings become significant.

Examples of such reactions have been quoted, varying from the despair of Monod, to the jaunty dismissal described by Tom Bell, and the magisterial condemnation of Adam Smith. Many others can be found, and together with the argument given in (ii) they form the basis of a growing literature which questions our present technological practice. The argument in (i) adds strong logical support.

References

1. Jacques Monod (1970). *Chance and necessity* (translation 1971), pp. 158, 160, Collins.
2. Izaak Walton (1654, reprinted 1939). *The compleat angler*, p. 176, Everyman.
3. Thomas Hobbes (1651, reprinted 1937). *Leviathan*, p. 65, Everyman.
4. Reference 1, p. 30.
5. H. H. Rosenbrock (1979). The redirection of technology, *IFAC Symposium on 'Criteria for selecting appropriate technologies under different cultural, technical and social conditions'*, Bari, Italy, 21–23 May.
6. Reference 1, p. 159.
7. Descartes (1637, reprint undated). *Discours de la méthode*, p. 109, Routledge (present author's translation).

8. P. Gaskell (1833). *The manufacturing population of England*, p. 183, Baldwin and Cradock.

9. See, for example, the many publications of Georges Friedmann.

10. Adam Smith (1776, reprinted 1976), *The wealth of nations*, V.i.f. 50, pp. 781–2, Glasgow Edition.

11. Thomas Bell (1941). *Pioneering days*, pp. 72–3, Lawrence and Wishart.

12. M. H. Pappworth (1967), *Human guinea pigs*, p. xi, Routledge and Kegan Paul.

13. Reference 12, p. 11. Pappworth is quoting Dr Szent-Gyorgi.

14. The account is based on conversations with a member of the research team.

15. Reference 12, p. 61.

16. Reference 12, pp. 92–3.

17. Lord Russell of Liverpool (1979). *The scourge of the swastika*, p. 164, Corgi Books; see also Reference 18.

18. Alexander Mitscherlich and Fred Mielke (1949). *Doctors of infamy*, p. 152, Henry Schuman, New York.

19. *Genesis*, Book 1, verses 29–30.

20. See, for example, Harry Braverman (1974). *Labour and monopoly capital*, Monthly Review Press.

21. *Work and the future* (1979). p. 20, CIO Publishing.

22. Frederick Winslow Taylor (1906). *On the art of cutting metals*, p. 55, American Society of Mechanical Engineers.

23. V. I. Lenin (1918). *Collected Works*, February–July, vol. 27, p. 259, Lawrence and Wishart.

24. Mammo Muchie, private communication,

25. Peter Brödner (1989). In search of the computer-aided craftsman, *AI and Society*, vol. 3, pp. 33–46.

8 Scientific management

A case in point

1. *Industrial production*

The industrial revolution, and the continuing development of industry that followed, provide a rich environment in which to study the impact of scientific values on society.[1] From the beginning, there were strong links between science and technology.[2] Wedgwood and Boulton were outstanding examples of the entrepreneur active as a scientist, but those who were known as scientists—Davy, Priestley and many others—also had strong ties with industry. So although technical developments at first arose much more from the mechanic and inventor than the scientist, there was always a community of ideas.

In the late nineteenth century, a new phase of development began, with the growth of the large corporation. Professional managers overcame the strong prejudice against them which had existed earlier,[3] and largely replaced the owner–manager. Science also began to be incorporated into industrial activity in a formal way, with the growth of industrial research laboratories. Factory organization continued the transition from the foreman as subcontractor to the integrated mass-production plant.

Still another rapid development is under way at the present time, driven by microelectronics and its applications to communications and computing. Any account of technology and of industrial organization would have to recognize these profound changes, and beware of generalizing about things so different as Boulton and Watt's Soho Works, and Henry Ford's production line, and a modern computer-controlled plant. Nevertheless, if we look at the assumptions which underlay the development of these very different kinds of production systems, we do not see the same sharp contrasts. Babbage,[4] and F. W. Taylor,[5] and a modern production engineer, present no such contrast in their ways of thinking.

2. *F. W. Taylor*

It is convenient to begin an examination of this enduring tradition with Frederick Winslow Taylor, who made himself its best-known exponent.

Taylor was born in 1856 into a wealthy Philadelphia family, and enjoyed the advantages which this position gave him. Highly competitive, he was an outstanding student, and with Clarence Clark won the U.S. Lawn Tennis Association doubles championship in 1881.[6] Though he passed the entrance examination to Harvard in 1874, he did not attend, but studied later through a curious extra-mural arrangement with Stevens Institute of Technology, and obtained a degree.

Instead of becoming a student at Harvard, Taylor began an apprenticeship as a machinist and pattern-maker, which he completed in 1878. From then on, he earned a good, but not apparently an exceptionally profitable living as an engineer with several companies, rising to chief engineer of the Midvale Company by the age of 28. During most of his industrial career, Taylor carried on research in technology, usually having a small team of helpers. He also developed his ideas of 'scientific management', and in 1893 became an independent consultant to propagate these ideas. After a precarious existence during five years of depression, he became consultant to Bethlehem Steel in 1898. In 1901 his consultancy terminated, and he retired with a substantial fortune which he claimed to have earned entirely (except for a $10 000 legacy) by his own efforts;[7] though he seems to have included among his efforts the management of his large investments.

During his career, Taylor had made a number of inventions, of which one—a tool-grinding machine—had a modest commercial success. With the assistance of Maunsel White, the metallurgist at Bethlehem Steel, he discovered 'high-speed steel'. This was not a new steel, but rather a process for hardening alloy tool steels at a higher temperature than normal. Sale of the rights brought him $50 000, and the publicity surrounding it made him one of the best-known engineers of his day.

If his career had ended at that point, Taylor would have been remembered as a highly successful engineer with controversial, and not markedly successful, ideas on industrial management. But from 1901 until his death in 1915, he propagated his ideas vigorously with the support of a small band of disciples. The controversy and propaganda during this period turned him into a national figure.

During a large part of his life, until he became engaged in controversy with the Trade Unions, Taylor had the good fortune to be taken at his own word and at his own estimate, and he mined a rich seam of fantasy. His largely sympathetic biographer admits that

> Taylor altered or expanded his accounts whenever it suited his immediate purposes. Though all of the stories in the Principles of Scientific Management had a factual basis, they were, with the possible exception of the Simonds episode, approximations of the actual events.[8]

And on Taylor's claims he remarks,

Of the four specific principles or achievements of scientific management that Taylor cited, the first was of questionable validity, the second was patently false, the third misleading, and the fourth, though suggesting the actual thrust of scientific management, magnified the effect of the third.[9]

3. *Scientific management defined*

Galileo held that 'it is the function of wise expositors to seek out the true sense,'[10] and nowhere is this more necessary than with Taylor. He contrasts scientific management, for example, with the traditional system that he calls 'initiative and incentive', in which 'the workmen give their best *initiative* and in return receive some *special incentive* from their employers.'[11] Under scientific management, on the other hand, management assumes new duties:

The new duties are grouped under four heads.

First They develop a science for each element of a man's work, which replaces the old rule-of-thumb method.

Second They scientifically select and then train, teach, and develop the workman . . .

Third They heartily cooperate with the men so as to insure all of the work being done in accordance with the principles of the science which has been developed.

Fourth There is an almost equal division of the work and the responsibility between the management and the workmen.[12]

With most writers, such phrases as 'develop a science,' 'scientifically select,' 'heartily cooperate,' and 'equal division,' could be interpreted in their common, domestic or everyday acceptation. With Taylor, it is necessary to treat them as terms of art, and to search out the sense in which he uses them. Fortunately he describes some of his investigations.

He says, for example,

The pig-iron handler stoops down, picks up a pig weighing about 92 pounds, walks a few feet or yards and then drops it on to the ground or upon a pile. This work is so crude and elementary in its nature that the writer believes that it would be possible to train an intelligent gorilla to become a more efficient pig-iron handler than any man can be. Yet it will be shown that the science of handling pig iron is so great and amounts to so much that it is impossible for the man who is best suited to this type of work to understand the principles of this science . . .[13]

The 'science' is described in the following way by Nelson.

Gillespie and Wolle selected ten of the 'very best men' and ordered them to load a car 'at their maximum speed'. Working at that rate, each man loaded the equivalent of seventy-five tons per day, nearly six times the previous average of thirteen tons. The men, however, were exhausted after filling only one car. Additional observations confirmed the conclusion that seventy-five tons per day was the theoretical limit. 'From this amount', Gillespie and Wolle reported, 'we deducted 40 per cent for rests and necessary delays, and set the amount to be loaded by a first class man at 45 tons per day.'[14]

Nelson comments that no justification was given for choosing the figure of 40 per cent. So far there is little that could be called science, but let us return to the story as told by Taylor.

We found that this gang were loading on the average about 12½ long tons per day. We were surprised to find, after studying the matter, that a first-class pig-iron handler ought to handle between 47 and 48 long tons per day, instead of 12½ tons. This task seemed to us so very large that we were obliged to go over our work several times before we were absolutely sure that we were right.[15]

The gain in scientific appearance, if not in content, is remarkable, and is achieved with the minimum of effort.

Taylor further justifies his increase in the daily task by invoking his 'law of heavy labouring,'[16] which was not invented until many months later, and rests upon equally shaky evidence. The 'law' was that 'for every load that a man carries on his arms he must be free from load for a certain percentage of the day and under load only a certain percentage. That is to say, a man carrying a ninety pound pig can only be under load 42 per cent of the day. He has to rest 58 per cent of the day.'[17] The 'science' in this is contributed by Taylor, who transforms the highly dubious data[18] into a precise and convincing 42%—or, since any figure would serve, 43% on another page.[19] The 58% (or 57%) of time not under load was of course made up largely of the time to return for another pig.

After the 'development of a science,' comes scientific selection, training, teaching and development of the workman. Selection was made after observation and consulting records, and it settled upon 'a little Pennsylvania Dutchman.' 'The task before us, then,' says Taylor, 'narrowed itself down to getting Schmidt to handle 47 tons of pig iron per day and making him glad to do it.'[20] The conversation which Taylor reports for this purpose became notorious, but seems not to have embarrassed him. It begins,

'Schmidt, are you a high-priced man?'
'Vell, I don't know vat you mean.'
'Oh yes you do. What I want to know is whether you are a high-priced man or not.'
'Vell, I don't know vat you mean.'
'Oh come now, you answer my questions. What I want to find out is whether you are a high-priced man or one of these cheap fellows here. What I want to find out is whether you want to earn $1.85 a day or whether you are satisfied with $1.15, just the same as all these cheap fellows are getting.'[21]

It ends in Schmidt's agreement to load 47 tons for $1.85 instead of 12½ tons for $1.15.* Taylor comments

* The facts, as usual, seem to have been somewhat different.[22] 'Schmidt', who was Henry Noll, was first offered $1.68 a day for loading 45 tons, and in fact averaged from $1.35 to $1.70 a day, apparently on slightly improved terms.

This seems to be rather rough talk . . . With a man of the mentally sluggish type of Schmidt it is appropriate and not unkind, since it is effective in fixing his attention on the high wages which he wants and away from what, if it were called to his attention, he probably would consider impossibly hard work.[21]

Training, teaching and development consisted of several days when

Schmidt started to work, and all day long, and at regular intervals, was told by the man who stood over him with a stop watch 'Now pick up a pig and walk. Now sit down and rest. Now walk—now rest,' etc. . . And he practically never failed to work at this pace and do the task that was set him during the three years that the writer was at Bethlehem.[23]

Hearty co-operation, under Taylor's third head, can be illustrated from his reorganization of the inspectors at the Simonds Company, which manufactured balls for cycle bearings. Nelson tells how Taylor assigned a special worker to help the inspectors maintain their pace:

In 1903 he described this person as an 'assistant' to the foreman, in 1911 as a 'teacher'. Each worker's output was recorded every hour, 'and they were all informed whether they were keeping up with their tasks, or how far they had fallen short'. If the deficiency was serious, the 'assistant' or 'teacher' was despatched 'to find out what was wrong, to straighten her out, and to encourage and help her to catch up.' Coupled with the 'scientific selection' of the workers [by which the slower girls had been sacked] this tactic must have been a powerful stimulus to maintain production. A visit from the 'assistant' or 'teacher' was a sure warning that the individual would be 'selected' for discharge.[24]

Taylor's fourth head called for 'almost equal division of the work and the responsibility between the management and the workman'. What he meant by this is that all responsibility for working methods was to be taken by management, and workers were to behave exactly as they were instructed:

Under our system the workman is told minutely just what he is to do and how he is to do it; and any improvement which he makes upon the orders given to him is fatal to success.[25]

4. *Scientific management evaluated*

Evaluating all this, we can say that Taylor's own account was largely a sham. Nelson assesses *The Principles of Scientific Management* in these terms:

It was inaccurate in historical detail and, more seriously, in its assertions about the nature of Taylor's contribution and the character of scientific management. Taylor's 'principles' had little or no relevance to his work, and his suggestion that time study constituted a 'science' was hyperbole at best.[26]

Nevertheless, Taylor's writings had immense influence. Nelson suggests that this was 'because he embraced the spirit and methodology of the modern scientific investigator and the large bureaucratic organisation',[27] but this suggestion needs deeper examination.

Taylor's model of physical science was one that was common enough in his time, though one that would not be accepted now. The investigator accumulated data, and on this basis attempted an inductive generalization. In practice, with Taylor, this was limited to the fitting of curves or equations to the data. In his investigation of cutting technology, for example, he spent 26 years and 200 000 dollars, according to his own account, in machining about 360 tons of steel into chips, varying more or less systematically the different parameters. Then he, or rather his assistants, fitted mathematical formulae to his data. The result was valuable, and 'Taylor's equation' in a modified form is still in use. There was, however, no real increase in theoretical understanding.

Taylor's equation describes the behaviour of a cutting tool in terms of efficient causes: 'cause and effect'. The original element in his work was the claim to be able to apply similar methods and the same kind of description to people.

> The contrast between Brinley and Davenport's rational, scientific approach to technical problems, and the foreman's capricious, often destructive approach to production problems, including the worker's behaviour, made a strong impression on him. There must be a better method, he reasoned, a way to apply the engineer's scientific perspective to management, as well as machinery. Taylor therefore arrived at the juncture many other engineers would reach in the latter years of the nineteenth century.[28]

One result of this attempt was Taylor's bonus system.[29] What he and many others were seeking was a formula relating pay to effort which would automatically cause the worker to produce at the maximum rate. It would be possible to think of this as the pursuit by the worker of some purpose, but it is clear that Taylor did not. Workers were to be shown, day by day, and even hour by hour, what work they had accomplished and so what wages they had earned. This cause would produce the appropriate effect.

Similarly, time study was to find the shortest time in which a task could be accomplished, and so to establish 'what constituted a day's work'. In regard to physical work, 'what we hoped ultimately to determine was . . . how many foot-pounds of work a man could do in a day'.[30] This search was barren, but Taylor claimed that he had found the right formulation of the problem, and its solution, in his pretentious 'law of heavy labour'.

Now baldly stated, in this way, Taylor's outlook seems simplistic and naive, but it is worth while to consider the justification which he could have made, and which, though not expressed, undoubtedly operated upon his thoughts. To the orthodox scientist, the only kind of explanation that is permitted is in terms of efficient causes. To the extent that anything can be understood scientifically, it must be possible to explain it in terms of 'cause and effect', or (in the usual way of thinking of machines) as a machine: 'In science man is a machine; or if he is not then he is nothing at all.'[31] It was naive of Taylor to believe that human behaviour could be explained in terms that were no more complicated (indeed simpler) than he had used to describe the operation of a machine tool. But in looking for a causal description of human behaviour, similar to the engineering description of a machine, he was faithfully carrying out the orthodox scientific programme.

To pursue the point, Taylor's engineering approach to the description of human behaviour was a dead end. The endeavour to provide a scientific explanation thereafter was taken up by social scientists, notably by Mayo,[32] and later by others. Now social scientists, to the extent that they are orthodox scientists, must seek for causal explanations. Then, if one has a causal explanation of human behaviour, it can be used to manipulate that behaviour: if a given cause has a given effect and we wish to produce the effect, we have the means at hand to do so. It is therefore not surprising that Mayo's work, like Taylor's, was criticized for being manipulative.[33]

An alternative view, in line with our earlier development, would see the possibility of a different explanation of human behaviour: in terms of purpose, following from a purposive development of physics. This would be equivalent in scientific terms to a causal explanation, being supported by the same evidence and refuted by the same evidence. But its effect upon our outlook and behaviour would be very different, just as the outlook of men who inhabit a Copernican universe will be very different from the outlook of those who live in a Ptolemaic world. In particular, a purposive view of human behaviour would free us from the compulsion to divide the world into observer and observed, manipulator and manipulated. Taylor's workmen would no longer be automata to be programmed into performing their tasks, but would have equal human status with himself.

We can imagine that to a man such as Taylor, highly competitive and unable to accommodate himself easily to the competitiveness of others, this alternative view would have been unattractive. He had struggled, as gang-boss of a machine shop, to impose his will on the machinists, and 'No one who has not had this experience can have an idea of the bitterness which is gradually developed in such a struggle.'[34] If he could explain the machinists' behaviour causally, and produce the effect he desired by manipulating the causes, none of this struggle or bitterness need arise: 'scientific management . . . may be summarized as:

i. Science, not rule of thumb.
ii. Harmony, not discord.
iii. Co-operation, not individualism.
iv. Maximum output, in place of restricted output.[35]

The fact that in adopting this approach he was treating others as not equally human with himself does not seem to have occurred to Taylor.

5. *Charles Babbage*

Braverman[36] suggests that Taylor must have known of the work of Charles Babbage, though nowhere referring to him. Babbage was a man of much higher intellectual ability than Taylor. He was Lucasian professor of mathematics at Cambridge from 1828 to 1838, and made contributions in a number of mathematical areas.[37] He conceived the idea of a stored-program digital computer, in principle equivalent to early modern computers. The story of his attempt to build a mechanical version of this machine, in which he spent much of his personal fortune as well as a government grant, is well known. He so impressed his contemporaries by his genius that his brain was removed after his death and examined, fruitlessly, to determine the cause of his superiority: it is still preserved by the Royal College of Surgeons. The macabre precedent was followed later for Lenin and for Einstein.[38]

Besides his scientific interests, Babbage followed the development of industry in Britain in the early nineteenth century, and was the first major analyst of its techniques.[39] His interest descended to the smallest particulars, as we can see from the following quotation.

> Two men are making an excavation, removing the earth in the usual way with spades and wheelbarrows.
>
> One of these men, Q., does more work than his companion P., and if an enquiry is made, Why is this so? the usual reply would be that Q. is either stronger, more active, or more skilful than P.
>
> Now it is the third of these qualifications which is the most important, because if Q. were inferior even both in strength and activity, he might yet by means of his skill perform a greater quantity of work without fatigue.
>
> He might have ascertained that a *given* weight of earth, raised at each shovelfull, together with a certain number of shovelfulls per hour, would be more advantageous for his strength than any other such combination.
>
> That a shovel of a certain weight, size, and form would fatigue him less than those of a different construction.[40]

Babbage continues in this way, enumerating four further variables in the operations of shoveling and barrowing. His account can be compared with Taylor's claim to have independently developed the 'science of shoveling':

> Now gentlemen, shoveling is a great science compared with pig-iron handling. I dare say that most of you gentlemen know that a good many pig-iron handlers can never learn to shovel right; the ordinary pig-iron handler is not the type of man well suited to shoveling. He is too stupid; there is too much mental strain, too much knack required of a shoveler for the pig-iron handler to take kindly to shoveling.

. . . Now if the problem were put up to any of you men to develop the science of shoveling as it was put up to us, that is, to a group of men who had deliberately set out to develop the science of doing all kinds of laboring work, where do you think you would begin? When you started to study the science of shoveling I make the assertion that you would be within two days—just as we were within two days—well on the way towards the development of the science of shoveling . . . I do not want to go into all the details of shoveling, but I will give you some of the elements, one or two of the most important elements of the science of shoveling; that is, the elements that reach further and have more serious consequences than any other. Probably the most important element in the science of shoveling is this: There must be some shovel load at which a first-class shoveler will do his biggest day's work.[41]

The account continues in a similar vein for sixteen pages, describing experiments with different types of shovel, and the organization by which the shovel to be used for each job was specified. It seems legitimate to suggest that Taylor could have saved himself his preliminary two days of thinking by reading Babbage, and may indeed have done so. In that case, his lack of acknowledgement was not without precedent.[42]

One difference between Babbage and Taylor is immediately evident from the above quotations. Babbage imagines the labourer himself to have the knowledge on which his skill is based:

Q. must, when a boy have been taught to examine *separately* the consequences of any defect or inconvenience in the parts of the tools he was to use in after life, or in the modes of using them. If not so taught he must have arrived at the same knowledge by the slower and more painful effort of his own reflections.[43]

For Taylor, on the other hand,

in almost all of the mechanic arts the science which underlies each act of each workman is so great and amounts to so much that the workman who is best suited to actually doing the work is incapable of fully understanding the science . . .[44]

The difference reflects the different organization of industry in the two periods, but also the difference in personalities. Babbage was a 'moderate radical,'[45] which in his day meant support of reform, of free trade, and of science and industry; and also for Babbage of workers' co-operatives.[46] He was a man of wide friendships, and generous to the workers he employed on his computing engines. Though an acute observer, he seems to have lacked managerial ability, and was unable to deal with the acquisitive opportunism of Clement, who acted as his subcontractor in the building of the difference engine.[47] It also has to be said that Babbage on many occasions describes with approval methods of subdividing labour which reduced the worker to the unskilled performance of procedures laid down by others.

6. *Claims of scientific management*

From Babbage, supplemented by Taylor, we can make a list of the principles and the claims of what became known as scientific management.

i. The division of labour. This implies not simply the specialization into trades such as carpenter and blacksmith, but a minute subdivision of the tasks to be performed in manufacturing. The practice is undoubtedly very old. Adam Smith,[48] apparently following Diderot's *Encyclopédie*,[49] gave the example of pin-making, which is elaborated by Babbage[50] in what is evidently an eye-witness account. He describes a workshop in which four men, four women, and two children were engaged in the various operations: drawing the wire, straightening it, pointing the wire, twisting wire to form the heads (which were made separately at that time), attaching the heads, tinning the pins and inserting them in papers for sale. The advantages of such a minute division of the work, as listed by Babbage and amplified by Taylor and others are these:

(*a*) Less time is spent in learning, which will reduce the cost of labour.
(*b*) Less material will be wasted in learning, which will reduce the price of the product.
(*c*) Time is not lost by the worker in exchanging one tool for another, nor in interrupting one occupation by another.
(*d*) Frequent repetition of one process leads to increased rapidity. For Taylor and his followers, this increased rapidity was to be obtained by rigid adherence to methods which had been devised by management.
(*e*) Subdivision of the work suggests the development of tools and machinery to carry out the separated operations.
(*f*) To these, which he regards as the usually accepted benefits, Babbage adds that the manufacturer, by subdividing the work, can employ skilled workers wholly on skilled work, leaving less-skilled operations to be done by others with less skill at lower wages:

> the master manufacturer, by dividing the work to be executed into different processes, each requiring different degrees of skill and force, can purchase exactly that precise quantity of both which is necessary for each process; whereas if the whole work were executed by one workman, that person must possess sufficient skill to perform the most difficult, and sufficient strength to execute the most laborious, of the operations into which the art is divided.[51]

ii. Separation of planning and execution. This extends (i), and Babbage[52] gives as an example Prony's organization of the computation of mathematical tables. A 'first section' contained five or six eminent mathematicians, who laid down the computational methods to be used. A second section comprised 'seven or eight persons of considerable acquaintance with mathematics', who reduced the formulae to numerical operations. The third section contained sixty to eighty people, who performed no other operation except addition and subtraction, according to the instructions and data they received. Taylor similarly removed all planning activity from the shop floor to a planning office, which issued detailed instructions for each operation. Planning entailed standardization of tools, cutting conditions, and working methods on lines laid down by Taylor.

iii. Time and motion study. This was foreshadowed by Babbage, but was mainly developed by Gilbreth and Taylor[53] and their followers. Motion study was intended to find the best and quickest way of doing work. Time study gave the time for a job by combining the times for separate motions, or alternatively by timing a worker.

iv. Payment by piecework of some kind. Taylor devised his own payment scheme,[54] in competition with many others. Workers were given a financial incentive to produce more.

v. Machine pacing. A later substitute for (iv), especially in assembly work, was to regulate the pace of work by the speed of the machines.

7. *Evaluation of the claims*

Examination of this list discloses a number of common themes. Some of the practices increase productivity by allowing more goods to be produced with the same effort. This is often true when new machines are developed, as for example a foot-operated hammer described by Babbage[55] for fixing the heads of pins. It has to be borne in mind that the account given above emphasizes the managerial aspects and under-emphasizes the effect of machinery. The development of 'scientific management' has been accompanied by a continued and accelerating development of machines.

A second theme, strongly emphasized by Braverman,[56] is the contention between employer and worker over the contract of employment. This is imprecise by its nature. The worker does not contract to produce a certain quantity of goods at a certain price: that would give the right to retain any profit that could be made on the bargain by better methods of working. Instead, the worker's labour is placed at the disposal of the employer, but not unreservedly.[57] What is expected will be partly set out in a formal contract, but this can never specify exactly what is to be done: attempts at closer definition may be self-defeating through 'working to rule'. Partly, what is required is a matter of custom, but there is much room for dispute in its interpretation.

The conflict of interest can exist not only under a capitalist, but also under a socialist organization,[58] or in some types of worker co-operative.[59] Each side will endeavour to re-interpret the terms of the contract to its own advantage, subject to the constraints of custom and law. That the problem is not a new one can be seen from the contract for the employment of masons at York Minster, in the second half of the fourteenth century, which is quoted by Frankl:[60]

> It is ordained by the Chapter of the church of Saint Peter of York that all the masons who shall work at the works of the same church of Saint Peter, shall, from Michaelmas day until the frst Sunday of Lent, be each day at morning at their work, in the lodge . . . as early as they may skillfully see by daylight in order to work: and they shall stand there truly working at their work all the day after as long as they may skillfully see to work . . . and at all other time of the year they may dine before noon, if they wish, and also eat at noon where they like, so they shall not remain away from their works in the aforesaid lodge no time of the year at dinner time except so short a time that no skillfull man shall find fault in their remaining away . . . and from the first Sunday of Lent until Michaelmas they shall be in the aforesaid lodge at their work at sunrise, and stand there truly and busily working upon the aforesaid work of the Church all the day, until there is no more space than time of a mile away before sunset . . . and if any man remain away from the lodge and from the aforesaid work, or make default any time of the year against this aforesaid ordinance, he shall be chastised with abating of his payment . . . and whoever comes against this ordinance and breaks it against the will of the aforesaid Chapter has he God's curse and Saint Peter's.

Many aspects of 'scientific management' can be seen as attempts either to specify the contract more closely, or to re-interpret it in the employer's interest. Taylor believed that by time study he could define exactly and objectively what was a 'fair day's work', and so avoid all dispute. But by demanding a higher rate of working than was customary, he was changing the accepted interpretation of the contract of employment in his favour.

A less obvious example is Babbage's principle that subdivision of work allows just that amount of skill and strength to be purchased which is needed for the work. If it has been customary for skilled workers to carry out a range of activities, with varying levels of skill, an employer may rearrange the work, giving the unskilled parts to workers at a lower wage. Suppose,[61] for example, that four cabinet makers were employed at £4 an hour, and spent a quarter of their time on the most demanding work, and three-quarters on less demanding work. An employer who engaged one cabinet maker at £4 an hour, and three workers of lower skill at £2 an hour, would save £6 an hour.

This, however, has been achieved by increasing fourfold the proportion of time spent by the remaining cabinet maker on the most skilled work. If he were selling his services on a professional basis (as, for example, a solicitor might do) he could argue that under the original conditions he was charging £10 an hour for one quarter of his time, during which he exercised superior skill and care, and £2 an hour for the remaining three-quarters.

Under the revised conditions he could claim entitlement to £10 an hour, and with three workers at £2 an hour the total wage would remain unchanged. An increased intensity of more-demanding work can equally be produced by changes in capital equipment, as at present in CAD systems.[62]

Noble[63] emphasizes a different but related aspect of the Tayloristic development of management and machinery, namely the quest for management control. By developments such as the machine pacing of work, and by withdrawing initiative and discretion from the worker, management can determine more completely the outcome of the production process. Again it seems that the same incentive can operate upon management whether the system is capitalist or socialist, or in some kinds of workers' co-operative.

If, on the other hand, one listens to a modern advocate of production engineering, most of which is strongly Taylorist, what is stressed is something different. The emphasis is upon economy and efficiency: a production-control system reduces work in progress, and thereby costs of production; so also does the linking of numerically controlled (NC) machine tools into a flexible manufacturing system (FMS) with automatic transfer of parts. A computer-aided design (CAD) system will allow iterations more quickly, to produce a better design. From the CAD system, programs can be produced automatically for the NC machines, allowing a rapid response to orders.

The whole production system, with its machines, its computer software, and its workers, is viewed in the scientific spirit as a 'machine': a system governed by equations which express 'cause and effect'. The system has been designed to fulfil a human purpose, and is to be judged by the efficiency with which it does so, but, in itself, it is thought to incorporate and express no purpose. It is to this exclusion of purpose, and to the treatment of the production system as a causal machine, that we shall trace many of the anti-human tendencies in Taylorism which have been described.

8. *Causality and purpose resumed*

In order to design a production system of the type described, one starts with the human purpose which it is to fulfil, just as one does in building a rocket with its guidance system. As in that case, the first step is to convert the purpose into relations of 'cause and effect': it was pointed out in Chapter 5 that we have no systematic way of constructing machines, or systems containing machines and people, without taking this preliminary step. Orthodox science describes anything which is capable of scientific description in terms of efficient causes. Our technology,with its scientific basis, will allow us to construct systems only when they are described in this way.

From this arises a profound difficulty. In its origins, work for men and women is the pre-eminent expression of purpose: we work to achieve an end. In a hunting and gathering community, work is hardly divided off from the rest of life, but it is what is done to obtain game or roots or berries, to prepare and cook them, to provide shelter and to meet the other needs of life. In a peasant community, work means tilling the soil, sowing and tending and harvesting the crops. The blacksmith and carpenter and wheelwright had an immediate purpose in making, and a more remote purpose in selling, the goods they made. Yet any human purpose in the operation of a modern production system is excluded by the terms in which we have to think of it.

When workers are included in such a system, it is as temporary substitutes for machines which we have not yet found it feasible or economic to devise. Men and women, in this view, have certain capabilities which it is difficult to emulate with machines: for example dexterity, mobility, and co-ordination of hand and eye. On the other hand, they are defective in speed, power, consistency and regularity. Developments in robot technology may allow us in time to match the human performance and exceed it, though, to quote an expert on robotics,

> it is less obvious that robots will be needed to take the place of human beings in most everyday jobs in industry . . . To bring in a universal robot would mean using a machine with many abilities to do a single job that may require only one ability.[64]

The way in which this view equates people to machines, and denies purpose to both equally, is clear. There is no room for human purpose within the framework of thought in which the system is conceived. The work performed by people in these systems is therefore denuded of all purpose that would give it meaning in itself. The only motivation that is left is that of payment, and this too is conceived to operate though 'cause and effect': two piece-rate systems, for example, might be compared experimentally in order to see which was the more effective in eliciting effort.

Once people have come to be regarded in this way as devoid of purpose, and as equivalent to machines, many of the features of Taylorism follow in a natural way. Like a computer, a human being is envisaged as needing a program before it can fulfil its function. Taylor supplied the program by means of the detailed instructions which he issued for the performance of every task.

A computer which failed to obey its program would be faulty, and would no longer be an effective device to fulfil its function. Hence Taylor's otherwise paradoxical statement that 'any improvement which [the worker] makes upon the orders given to him is fatal to success'. An improvement, one would think, is still an improvement, even if unlooked-for. But if the worker-computer fails to obey its program, all hope of using it effectively is lost.

Human beings who are to fulfil the role of programmable devices must be standardized, because otherwise a different program will be needed for each worker. But as people differ in their skill and experience and ability, none of this can be drawn upon. The standard man or woman for whom tasks are designed must have no more ability or skill than can reliably be found in every worker. Necessarily, a machine with such elementary properties must be supplied only with the simplest programs, and tasks must be simplified so they can be learned in the shortest possible time. Ninety-five per cent of Henry Ford's workers[65] had to be 'skilled in exactly one operation which the most stupid man can learn within two days', 'The man who puts in a bolt does not put on the nut; the man who puts on the nut does not tighten it.'

Taken to its extreme, this attitude leads to the view that many industrial tasks are most suited to the abilities of the mentally handicapped.[66] Lacking some part of a full human capacity, they fit better to the specialized demands of modern production systems. One example will suffice to illustrate a pathological extreme of the Tayloristic trivialization of work.

> Four years ago, Roland Temme started his machine shop in Lincoln, Neb, with one \$1100 lathe. This year, that shop, called TMCO, will do \$250 000 in business, three-quarters of that on labor-only contracts.
>
> But perhaps even more important than the growth of the seven-employee shop is the way in which two young men, classified as mentally handicapped, have contributed to the company's success . . .
>
> Mike Bayless, 28 years old with a maximum intelligence level of a 12-year-old, has become the company's NC-machining-center operator because his limitations afford him the level of patience and persistence to carefully watch his machine and the work that it produces . . .
>
> Mike Bayless has been trained to know exactly what the Moog Hydra-Point machining center does with each table of parts that he loads into it. 'His big plus, though,' says Temme, 'is that he will watch the machine go through each operation step by step, and he doesn't hesitate to hit the "Stop" button if it doesn't look right.'
>
> That patience and watchfulness saved TMCO a bundle not too long ago, recalls Temme, because Bayless anticipated a problem before it could develop fully. 'If a so-called normal individual had been assigned to the machine, he would have been doing inspection work or deburring and would never have known what was happening until it was too late. Mike hit the button at the first hint of something wrong and saved us significant downtime and repair.'
>
> Mike's meticulous attention to the operation of the machining center has also saved TMCO rework and reject costs. 'He loads every table the way he has been taught, watches the Moog operate, and then unloads. It's the kind of tedious work that some non-handicapped people might have difficulty coping with,' Temme points out.[67]

9. *Summary*

To summarize the argument of this chapter, 'Scientific Management' represents an attitude which has been current since the beginning of the

Industrial Revolution, and is still predominant in technology. It has been refined and gradually developed, for example by the introduction of time and motion study, but its fundamental basis has remained unchanged. It is rejected by social scientists, but to the extent that they adopt causal explanations of human behaviour, their work fails to counteract the underlying assumption of Taylorism.

Taylor's explanation of scientific management is largely a smoke screen. When analysed, Taylorism has three main components:

i. Improvements in working, and in machines, which offer the benefit of greater production for an unchanged human effort. Taylorism takes credit for these improvements, even when they are independent of its other aspects and could be obtained without them.

ii. A number of techniques to allow the employer to redefine the contract of employment to his own advantage. This can increase the intensity of work and alter the distribution of benefits to the disadvantage of the worker. The just distribution of the rewards provided by modern technology is a political problem, and no existing system meets with full acceptance by all participants.

iii. The withdrawal of control and initiative from the lower levels of an organization, and their concentration, to the greatest possible extent, in the higher levels. This entails the simplification of tasks, and their precise definition, which assists in their subsequent mechanization.

One suggested explanation for this last tendency is that it arises from the effort to extend the control by management of the production system.[68] What has to be explained, however, is why control is sought by these particular means, and why this is done under all forms of organization: capitalist, socialist, and co-operative. There is no diversity in the approach to technology to correspond with the diversity of means which have been tried in order to achieve a just distribution of the material benefits which technology can bring. This uniformity is the more surprising because such small departures from it as do exist are often accompanied by marked increases in productivity.[69]

The explanation offered here is that the tendency arises from our acceptance of the causal myth. This is incorporated in science, and thence transferred to technology. Rational analysis and design then inevitably lead to a reductionist treatment of work, which is fragmented into the smallest parts. These can be specified in causal terms, as though to describe the operation of a machine. The fragmented tasks are given to individual workers to perform, and recombined as necessary to achieve the aims of production. The expression of purpose by a worker is rejected, or more accurately is excluded from all consideration because rational analysis in causal terms leaves no room for it.

References

1. See for example Arnold Toynbee (1884, reprinted 1969). *Lectures on the Industrial Revolution in England*, David and Charles Reprints; Paul Mantoux (1928). *The Industrial Revolution in the eighteenth century*, Jonathan Cape; E. P. Thompson (1963). *The making of the English working class*, Gollancz, (Penguin edition, 1968).
2. Robert E. Schofield (1963). *The Lunar Society of Birmingham*, Oxford, Clarendon Press.
3. Sidney Pollard (1965). *The genesis of modern management*, Edward Arnold.
4. Anthony Hyman (1982). *Charles Babbage, pioneer of the computer*, Oxford University Press.
5. Daniel Nelson (1980). *Frederick W. Taylor and the rise of scientific management*, University of Wisconsin Press.
6. Reference 5, p. 28.
7. Reference 5, p. 105.
8. Reference 5, p. 172.
9. Reference 5, p. 171.
10. Galileo (1615). Letter to the Grand Duchess Christina. In *Discoveries and opinions of Galileo*, translated by Stillman Drake (1957), p. 186, Doubleday.
11. Frederick Winslow Taylor (1911, reprinted 1967). *The principles of scientific management*, p. 34, Norton Library.
12. Reference 11, p. 36.
13. Reference 11, pp. 36–7.
14. Reference 5, p. 92.
15. Reference 11, p. 42.
16. Reference 11, pp. 53–8, 60–1; Reference 5, pp. 91–8.
17. Reference 5, p. 96.
18. C. D. Wrege and A. G. Perroni (1974). Taylor's pig-tale: a historical analysis of Frederick W. Taylor's pig-iron experiments, *Academy of Management Journal*, vol. 17, pp. 6–27.
19. Reference 11, p. 57.
20. Reference 11, p. 44.
21. Reference 11, pp. 44, 46.
22. Reference 5, pp. 92–5; p. 223, note 97.
23. Reference 11, p. 47.
24. Reference 5, p. 74.
25. Frederick Winslow Taylor (1906). *On the art of cutting metals* (3rd edition, revised, undated), p. 55, American Society of Mechanical Engineers.
26. Reference 5, p. 172.

27. Reference 5, pp. 198–9.

28. Reference 5, p. 34.

29. Frederick W. Taylor (1914). A piece rate system. In Clarence Bertrand Thompson (editor), *Scientific management*, pp. 636–83, Harvard University Press.

30. Reference 5, p. 43.

31. Joseph Needham (1927). *Man a machine*, p. 93, Kegan Paul.

32. Loren Baritz (1960, reprinted 1974). *The servants of power*, Greenwood Press.

33. Reference 32, passim, but especially pp. 96–116, 191–210.

34. Reference 11, p. 50.

35. Reference 11, p. 140.

36. Harry Braverman (1974). *Labor and monopoly capital*, p. 89, Monthly Review Press.

37. See Reference 4; also J. M. Dubbey (1978). *The mathematical work of Charles Babbage*, Cambridge University Press.

38. Christopher Evans (1979). *The mighty micro*, p. 28, Gollancz; Steven Rose, Leon J. Kamin, and R. C. Lewontin (1984). *Not in our genes*, p. 53, Penguin.

39. Charles Babbage (1832, enlarged 1835, reprinted 1963). *On the economy of machinery and manufactures*, Augustus M. Kelley.

40. Charles Babbage (1851). *The exposition of 1851*, pp. 4–5, John Murray.

41. F. W. Taylor (undated). Testimony before the Special House Committee, pp. 50–51. In F. W. Taylor, *Scientific Management*, Harper and Brothers, New York.

42. Reference 5, p. 54.

43. Reference 40, p. 5.

44. Reference 11, pp. 25–6.

45. Reference 4, p. 38.

46. Reference 39, pp. 253–9.

47. Reference 4, pp. 123–32.

48. Adam Smith (1776, Glasgow edition 1976). *The wealth of nations*, I.i.3, pp. 14–15.

49. Diderot et d'Alembert (1755). *Encyclopédie, ou dictionnaire raisonné des sciences, des arts et des métiers*, vol. 5, pp. 804–8, article 'épingle'.

50. Reference 39, pp. 176–87.

51. Reference 39, pp. 175–6.

52. Reference 39, pp. 191–6.

53. Reference 11, pp. 77–86.

54. Reference 29.

55. Reference 39, pp. 180–2.

56. Reference 36, pp. 54–8,

57. See Council for Science and Society (1981). *New technology: society, employment and skill*, pp. 19–20, CSS Report.

58. Miklós Haraszty (1977). *A worker in a worker's state*, Penguin Books: compare Göran Palm (1977). *The flight from work*, Cambridge University Press; Robert Linhart (1981). *The assembly line* (translated by Margaret Crosland), John Calder, London; Satoshi Kamata (1983). *Japan in the passing lane* (translated and edited by Tatsura Akimoto), George Allen and Unwin.

59. Alasdair Clayre (1980). *The political economy of co-operation and participation*, Oxford University Press; Robert Oakeshott (1978). *The case for workers' co-ops*, Routledge.

60. Paul Frankl (1960). *The Gothic*, pp. 852–3, Princeton University Press. The document is about 2½ times longer than this quotation and includes, for example, provision for sleep after the midday meal.

61. Reference 57, p. 14.

62. Mike Cooley (1987). *Architect or bee?*, Chatto and Windus.

63. David F. Noble (1977). *America by design. Science, technology and the rise of corporate capitalism*, Alfred A. Knopf.

64. F. H. George and J. D. Humphries (editors) (1974). *The robots are coming*, p. 164, NCC Publications.

65. Henry Ford, in collaboration with Samuel Crowther (1923). *My life and work*, pp. 87, 83, Heinemann.

66. See, e.g., Eikonix Corporation (1979). *Technology assessment. The impact of robots*, 30 September, p. 157, US Dept. of Commerce, National Technical Information Service PB80-142268; Georges Friedmann (1955). *Industrial society*, p. 216, Free Press of Glencoe; Frederick Herzberg (1966). *Work and the nature of man*, p. 39, World Publishing Co.; James O'Toole (1974). *Work and the quality of life. Resource papers for Work in America*, p. 16, MIT Press; A. D. Swain (1977). Design of industrial jobs a worker can and will do. In S. C. Brown and J. N. T. Martin (editors), *Human aspects of man-made systems*, p. 192, Open University Press.

67. *American Machinist* (July 1979). vol. 123, no. 7, p. 58.

68. See References 36, 63.

69. John E. Kelly (1978). A reappraisal of sociotechnical systems theory, *Human Relations*, vol. 31, pp. 1069–99.

9 An alternative technology

1. *Technology from a different viewpoint*

The account of Babbage's views, and still more those of Taylor, makes the point that our present technology incorporates values which come directly from science. All of nature is seen as a machine without purpose, though each person makes a lenient exception in his own interest. Then those who design any technological system are in general unable to see its machines and its workers in any other way than as causal devices, responding in fixed (if complicated) ways to the stimuli which are applied to them.

This outlook is seldom adopted in a consciously anti-human way. In social life and in everyday matters, people are treated with the courtesy and respect which society expects. But in technical matters they have to be treated as machines, as causal devices, because these are the only terms in which our science, and our science-based technology, can deal with the world.

We are so used to this situation that it seems inescapable, a fact of nature. All observation of nature, all scientific study of people, endorses this causal behaviour. Every phenomenon, when scientifically understood, is explained causally. Every object, living or dead, is enmeshed in an endless chain of cause and effect.

The origins of this apparent inevitability have already been explained. Science begins with an a priori decision that no explanation will be accepted unless it is in terms of cause and effect. Every success which we have in describing and explaining the world then gives us new examples to reinforce the belief that nature itself is intrinsically causal.

An alternative view is that nature knows nothing of causality, which is a structure that we ourselves impose upon the facts by means of our explanations. We could impose other structures, and in particular could explain the world in terms of purpose. Causal behaviour would then follow as a description of the policy which implemented the purpose. The causal and purposive explanations would be equivalent for every need of science, but not equal in their effect on our outlook and behaviour.

If, over the past four centuries, our science had given all explanations in terms of purpose, we should believe that nature itself was purposive. Every observation and every explanation would reinforce this view. It would then be legitimate, and perhaps useful, to point out that the idea of purpose was unnecessary. Every purposive explanation could be replaced by the causal explanation arising from the policy, just as any general, causal law in science could be replaced by an exhaustive list of special cases. There would, in either case, be a loss in mathematical coherence, but also a shift in outlook implying different values.

In other words, our aim has not been to replace one myth by another. It has been to show that there can indeed be equivalent myths, and that these carry values which affect our behaviour. As a case in illustration, Taylorism is a direct response to the causal myth. What we shall attempt in the present chapter is to show how the purposive myth allows us to regard technology in a different light. This will be done in the hope that the insight obtained from this alternative view will release us from the tyranny of the causal view, allowing us to suggest a better kind of technology, better matched to human needs and aspirations.

2. *The scope of technology*

The word 'technology' is sometimes restricted to mean the knowledge and skill which underlie our processes of production and distribution and their control. The physical embodiment of technology in systems containing people and machines is then distinguished as 'technics', though this term is not common in English writing. 'Technology' will be used here in a wide sense to embrace the conceptual and physical aspects. Where necessary, distinctions will be made by using such terms as 'technological system', 'plant', 'machine', 'skill', or 'scientific basis of technology'.

Technological systems may be of the most diverse kinds. A railway is a technological system, comprising lines, rolling stock, signalling equipment, maintenance facilities, and the people who operate, and organize the operation, of this physical equipment. A road haulage company or an airline are similar examples, but ones which make extensive use of independent systems: roads and airports.

Production systems are a major subdivision of technological systems. They may be based upon continuous flow with nearly complete automation, as in an oil refinery. Steel production is only a little less continuous and less automatic. Mass production of consumer goods follows the same pattern as far as possible: closely in the case of food, less closely in such goods as motor cars where final assembly employs much human labour. In all of these systems, computers will usually be found, sharing with people the tasks of control.

Other systems are concerned with providing services of various kinds: shops, banks, insurance offices, hospitals and the like. The military services, army, navy and air force, are technological systems comprising people, machines, communications systems, etc. Government offices are another type of technological system.

A common feature is that specialized organization, or specialized equipment, is always involved, usually with human assistance but sometimes without. So we should not call an individual lecturer and his audience a technological system; but a university is a technological system through its specialized organization and equipment. Traffic lights, and a completely automated telephone system, are both technological systems, which operate without human intervention.

What is, or is not, included in the term 'technological system' clearly cannot be made entirely unambiguous, but one distinguishing feature serves as a useful guide. This is, that underlying the activity which is carried out, there is some feature which is amenable to study in science-based terms: that is, in terms of technology in its abstract meaning.

A telephone system, for example, is designed using mathematical techniques such as queuing theory, and is supported by highly developed theories of electronics, computing, and optical transmission of signals. There are specialized techniques for installation and maintenance, and a specialized organization for billing. All of these are based upon the aplication of science in technology; and the scientific basis, as in other technological areas, becomes continually more refined and more powerful.

3. *The purpose of technology*

Every technological system starts from a human purpose, from the intention to satisfy some human need or desire. The intention may arise from the aims and actions of government, as nearly always in police or military systems, and sometimes also in many other areas. Alternatively, the intention may arise from the wish to meet a market need, in which case it is subordinate to other aims such as immediate profit for a company, or the fulfilment of a longer-term aim of 'profitable survival'.

However it is formed, the purpose is implemented by a process similar to the one illustrated in Figure 5.9. That is to say, a policy which will implement the purpose is generated, and the system is constructed so that it follows the causal laws implied by the policy. This procedure is much less formal in most cases than in the control engineering example of Chapter 5. To begin with, the purpose is seldom defined so precisely that a unique policy can be deduced from it. Then the procedure by which a policy is obtained must also be less rigorous, and it usually involves intuitive elements of design and problem-solving.

Nevertheless, a little thought will show that the way in which the technological system is defined follows closely the outline given before. We have certain ends in view, and we generate the means to obtain them. These means will consist of organized groups of people, of machines, and of interactions between people and machines. All of these, if they are to be brought within the scope of scientific study, conceived in the traditional way, must be described causally, because the only science we have is cast in the mould of the causal myth.

If we replace the causal myth by the purposive myth, we at once have to change the way in which we regard the technological system. The people involved in the operation of the system will be seen as sharing in the purpose of the system, and assisting in its fulfilment. Individually, they will have purposes which are subordinate to the purpose of the system; that is, their independent purposes which they strive to fulfil will be consistent with the purpose of the system, but not coextensive with it. The human contribution to the operation of the system will consist of an interlocking network of these subordinate purposes.

The machines will also embody subordinate purposes, but in a more restricted way. Human beings can adopt a purpose at a deep level: we say that they care whether the purpose is fulfilled or not. Because of the limitations of our technology, we cannot implement purposes in machines at this level. Machines do not care whether they fulfil their purpose. We can make a machine behave 'as if it cared', within a restricted repertoire, by making it exhibit the causal behaviour which would go with caring. But this behaviour is superimposed upon an alien purpose (expressed by Hamilton's principle and the like) out of which it does not grow in a natural way. Human purposes can equally be imposed by coercion in an alien way, but they can also be adopted freely.

Within such a purposive view of technology, it would be necessary to consider the interactions of various kinds: of people with people, of machines with machines, and of people with machines. Each of these three kinds of interaction will be considered briefly.

4. *Interactions between people*

To regard the interactions between people in terms of their individual purposes is not something new. This is the natural view, expressed in most of our literature and our conversation. The alternative, causal, view is less natural and of later growth: an early and crude form was briskly dismissed by Socrates (Chapter 3, Section 6) but it is in the past four hundred years that it has become most widespread.

A current view is that human behaviour can be explained partly by heredity and partly by the influence of our environment. The first can ultimately be deduced from the chemical properties of DNA, resting upon quantum mechanics. The second can be derived from structural properties of the brain, responding to physical inputs via sight, hearing, and other senses. This mechanical picture is somewhat softened by the suggestion that the complicated structure of the body requires new categories for its description; but these categories are equally causal, and would not resist an explanation in more detailed and basic terms.

It is this outlook which leads to Taylorism, and a manipulative view of human relations in technological systems. A purposive description avoids this danger, and situates the problems of personal relations in a more traditional framework. This does not mean that it solves the problems. It simply restores them to a moral dimension from which the causal account would remove them. Rather than observer and observed, manipulated and manipulator, it shows us moral beings of equal status co-operating or conflicting through their purposes.

This change is particularly relevant in two areas, of ownership and of work. In the causal view, there is a justification for ownership in an absolute sense. What is inert and without purpose can be owned by being made absolutely subject to our own will, which we exempt from the otherwise universal rejection of purpose. This conception of ownership applies to our relation with nature, to our ownership of land and plants and animals. It applies also to the ownership by one person of another person's time, in a contract of employment, to the extent of specifying what shall be done to the exclusion of all initiative and independent will.

If, on the other hand, our own purpose is only one in a meshing network of purposes, we cannot own absolutely. We can force nature to conform to a specified pattern of behaviour by a procedure resembling the one given in Section 10 of Chapter 5. But this is only a semblance of conformity; as we can make a machine 'seem to care', or a computer 'seem to be intelligent'. The underlying purpose in all cases has not been changed, and by escaping from our control it limits the extent to which we can 'own'.

The same applies to human relations. Through rewards and penalties, workers in a Tayloristic system can be brought to conform to a specified behaviour. But their underlying human purpose has not been changed, and the conformity is only superficial. All the well-known problems[1] will arise which come from a lack of commitment, and the imposed control will continually require to be tightened.

The purposive view leads in a different direction, towards a system in which workers share in the purpose which is to be attained. Their own part in this purpose, the fulfilling of a subordinate purpose, must be one which is acceptable and natural to them. It must give scope for the exercise of their own abilities and skills. They in return will have to accept responsibility for their own part in fulfilling the aims of the system.*

Nothing in this is very new. What is different is the justification which could be given, and the likelihood of its acceptance. To argue for such a system in the face of the causal myth constrains us strongly to accept the values which that myth implies. It has to be argued, for example, that a system with these features is more effective than a Tayloristic system in eliciting the desired behaviour from a worker. That is, the justification must be in causal terms; and once the causal view of human behaviour is admitted it brings us round again full circle to Taylorism.

5. *The nature of work*

The relevance of the purposive myth to the nature of work is even more direct than its relevance to ownership. Work in modern societies has three functions. It contributes to the production of goods and services. It gives a claim to some part of this production. It is also, for most, a major source of self-esteem through the contribution which it allows them to make to society: it is by breaking this link with society that unemployment can be so damaging.

As the direct contribution of human effort to production continually decreases, these three aspects of work become problematic. Our view of the future then depends very much on whether we adopt the causal or the purposive myth. The first of these suggests that since 'man is a machine', every human contribution to production can ultimately be eliminated by machines. Goods will be produced in the workerless factory. They will be distributed by ever more automated distribution systems, needing progressively less human input. Communication by electronic systems is already highly automated, and will become more so. Other services equally can be provided with less and less human labour.

However enthralling this development might be to the technologist, it is a bleak prospect in human terms. It seems that we shall become less and less necessary to the society in which we live, existing on the margins of a mechanized and automatically functioning world. Our contribution to this activity will hardly be needed, and when it is, the contribution will be as mechanical as the system it serves. Work as an entitlement to share in what is produced will become a source of contention through its scarcity. Work as a source of satisfaction and self-esteem will cease to exist.

* It may be that some workers, at some stage of their lives, might not wish to accept this kind of responsibility, or to accept it only in a small degree. This might be true particularly for those who are inured and accustomed to present types of work, or who have strong alternative interests. There would be no difficulty then in providing them with appropriate tasks, particularly if work was distributed among autonomous groups. But with an outlook formed by the purposive myth this would appear as anomalous, as a failure to accept the normal obligations of society: a failure which is to be tolerated if it is strongly desired, but not to be encouraged and never to be imposed.

Repelled by this prospect, some have suggested a reversion to earlier technology, or a breaking of the machines.[2] But the vision of the future springs directly from the causal myth, and a different future can be imagined if we adopt the purposive myth. This accepts that 'man is a machine', but a machine with a purpose. Those things which we usually call machines, including computers, also have a purpose, but it is very different from the purpose of human beings.

One part of our human purpose is to produce the goods and services we need. To some extent, and increasingly, we can incorporate this purpose in machines. But we can do so only by first translating it into a policy as explained in Section 10 of Chapter 5. The machine does not become more like the human being from whom the purpose originated; it merely reproduces a desired behaviour. It does so within a limited repertoire of actions, and in a limited range of environments. If pushed sufficiently far outside its repertoire or its range of environments, it will ultimately exhibit a purpose which diverges from the one which a human being would adopt.

By incorporating the human purpose in a machine, we have made this purpose subordinate to the ultimate purpose of the machine—that purpose expressed by Hamilton's principle and its extensions from which, as a policy, all its actions can be deduced. In the human being, the purpose was subordinate to a different ultimate purpose, one of flesh and blood, not one of rods and levers and silicon chips. So although we can expect to incorporate in a machine more and more human purposes, we can do so only in a restricted way.

The human actor 'fulfils a purpose' in the sense that his actions can be derived from this purpose, which is subordinate to an ultimate human purpose. But the actions occur without necessarily being derived explicitly and expressed by a policy; just as the Earth goes round the Sun without any regard to Hamilton's principle or to that policy derived from it which is expressed by Newton's equations. The purpose and the policy are our description of an event which simply occurs. In the same way, the machine fulfils its ultimate purpose naturally, but we have so constructed the machine that it has the human purpose as one of its subordinate purposes: but only in those circumstances where the purpose gives rise to one specific policy or range of policies. Because it is the policy which we impose on the machine, not the purpose.

Despite this restriction, machines in which we have incorporated a human purpose (in the sense defined) will be able to carry out a wide range of simple human tasks. Traffic signals can replace the police who at one time controlled traffic. Robots can put in spot welds on car bodies. Repetitive work of this kind makes no real use of human abilities, and if the time released can be used in more humanly rewarding ways, we can accept it as a benefit.

Those machines which are based upon computers can carry out more complex tasks. If we can, by questioning or observing a competent worker (the 'expert') define exactly what is needed to accomplish a task, we can build this information into the computer as an 'expert system'. Then the computer can carry out the task, however complicated it may be, if it is provided with the necessary devices for sensing and for manipulating its environment.

The tasks which can be most easily defined in this way are often ones which we regard as prime evidence of human intelligence: calculating numbers, manipulating mathematical formulae, playing chess at a moderately advanced level, some parts of medical diagnosis, and so on. When it is shown, as it has been,[3] that all of these can be done with more or less competence by a computer, we are apt to conclude that 'everything that man can do, a computer can do'. Computers have already shown the ability to do things we regard as difficult, and as indicative of high human abilities: why should they not also be able to perform every other task which is within human capability?

If we accept the causal myth, the conclusion is indeed inescapable, for every human activity is then defined by the actions it entails. But if we adopt the purposive myth we shall see much human activity as fulfilling a purpose directly, without first translating the purpose into the policy which will fulfil it.

We should then be led to see ourselves as distinguished from the machines by what unites us to the other animals: by the fact that our purposes, with the actions which result from them, are subordinate to the purpose of flesh and blood. We should see ourselves as distinguished from the rest of the animal world by what unites us to machines: the fact that we can implement complicated logical sequences of cause and effect which represent the policy needed to fulfil a purpose. We should see ourselves as distinguished from both, and unique, by our capacity to implement purposes which span these two kinds of activity.

On this view, mathematical calculation is not specifically human, because it can be done by machines. Bodily activities, such as jumping and running, which implement an untranslated purpose, are not specifically human, because they are shared by other animals. The generation of a new mathematical theorem, as the fulfilment of a purpose which is not first translated into causal terms, is a specifically human activity. So is the act, equally fulfilling an untranslated purpose, by which a reverberation is set up between words and their meaning:

Shall I compare thee to a summer's day?
Thou art more lovely and more temperate

or in a darker mood,

And this is the manner of the Daughters of Albion in their beauty.

So also is the activity of the painter or sculptor, and at a more accessible level the exercise of craft skills such as those of the carpenter or blacksmith, and the practice of engineering design.

All of these are activities which implement a human purpose, and do so without first translating it completely into the causal actions which will be needed. They are all activities, in consequence, which resist formalization and the reduction to method, but must rest on a basis of ability developed through experience. In the tradition of the causal myth, they are therefore ignored, or mentioned only briefly in passing. In Popper's account of the scientific method, for example, the process by which a new theory is generated is outside the logical framework, and is left undefined, though it is the most specifically human part of the whole activity.

What is most distinctive in human work, on this view, is that activity which fulfils a human purpose in its untranslated form. The purpose must involve more than the simple activities which we share with other animals, such as the co-ordination of hand and eye in muscular activity. It must be more than the logical, causal sequence of actions which we share with machines, and which Taylorism would see as the only content of work. Both of these may be components of a truly human kind of work, but there must also be that element of directedness, of aiming at a goal, and achieving it by judgement and skill based upon knowledge and experience.

In these terms we can imagine a different future for human work. Instead of the workerless factory and the marginalizing of all human contribution, we can imagine a technology in which the human contribution is central: where this contribution is assisted by machines, but where it retains the quality of directly fulfilling a purpose. A technology conceived in this spirit could provide a truly human kind of work, and could produce goods and services which met more amply the needs of those for whom they were provided.

The knowledge and the skill required by such a technology will be at different levels in different kinds of work, and will be appropriate to different kinds of human ability. But no human work should have the machine-like triviality and aimlessness of the lamp plant (Section 5 of Chapter 7). This offends by equating people to automata, fulfilling a purpose appropriate only to machines. It also offends in another way, by subordinating people to machines.

6. *Machines and people*

The distinction just made can be illustrated by Henry Ford's introduction of the assembly line. This was accompanied by an extreme subdivision of the work, expressed in his dictum,[4] 'The man who puts in a bolt does not put on the nut; the man who puts on the nut does not tighten it.' In such a working situation, all human content has been eliminated by an ultimate trivialization of the task. The work is machine-like, and men are treated as though they were machines. This would be true regardless of the other aspects of their work: for example if they were assembling components which they picked from a bin. In these circumstances the workers might be subject to pressure to increase the rate of production, and the situation could be unsatisfactory to varying degrees.

The production line, however, goes beyond this. When he introduced it, Henry Ford regarded it as a way of bringing the work to the worker, and eliminating the time otherwise lost in moving from one car to the next. Its effect is more profound. It links all the separate parts of the assembly process into one whole. The assembly plant becomes one vast machine, in which certain actions are indeed carried out by machines, while others, equally machine-like, are carried out by workers.

Here the whole system has a purpose, which was incorporated into it by its designers. They did this by incorporating a purpose into the machine components, and requiring workers to conform to the purpose of the machine. When the production line speeds up, workers must also speed up. When it slows down, they must also slow down. A car is presented to them, and they have so many seconds to carry out their task. The machine, the production line, defines what they are to do and how fast they are to do it. All control is taken from them, and they are required to subordinate themselves as servants to the purpose of the machine.

This inverts the traditional relationship, in which a craftsman may use machines, but they are subservient, as tools to assist him in accomplishing his purpose. The purposive myth suggests that this traditional relationship is the correct and humanly satisfactory one; but it is extraordinarily difficult to defend it or promote it within the framework of the causal myth.

A first difficulty is that the causal myth rejects purpose. It does not allow us to say that machines have a purpose, or that the purpose of a worker is subordinated to the purpose incorporated in a machine. Definition of the problem becomes impossible, because the terms needed to discuss it are disallowed.

Secondly, human purpose itself tends to disappear. An unsatisfactory working situation may lead to problems—to poor quality of work or to a high labour turnover. The difficulty may then be studied, and perhaps in a sympathetic spirit, in order to provide a more satisfactory working situation. But if this is done in a 'scientific' way, which means in the light of the causal myth, the problem has to be stated in the form, 'what changes in the working situation will cause it to appear better in the eyes of workers?' Even if the tendency to manipulation is resisted, the question will not lead to the reply, 'the machines should be made subordinate to the purpose expressed by workers in their tasks'. The attempt to design a more humanly satisfactory technology therefore meets a basic difficulty, because the terms in which it would have to be done contradict the orthodox understanding of science, upon which our technology is based.

7. *The relation of machine to machine*

Of the three types of relation which were described earlier, we have said that the relation between people is situated by the purposive myth in a traditional framework, pre-dating the causal myth and still persisting. Different people, and different groups of people, have purposes which sometimes coincide and sometimes conflict. Little can be said about this that is not already incorporated in the traditional account, which has always been expressed in terms of purpose.

The relation between people and machines, by contrast, cannot be treated in a satisfactory way without a change in the usual terms of discussion, which are based on the causal myth. It is this change which has been our main subject. The third item, the relation between machine and machine, offers no fundamentally new difficulty.

Machines can relate to one another in two ways, either by the passage from machine to machine of material for processing, or by the interchange of information. When passage of material is accomplished manually, the important interaction is between machines and people, and this has already been discussed. Where it takes place automatically, the linked machines become one larger machine, as in the car assembly line. The important question is again the relation between this larger machine and the people whose work it either controls or assists.

This leaves the flow of information as the new factor to be considered. There are two different extreme forms which this flow can take, with an unlimited number of intermediate forms. At one extreme, all information can be sent to a single, central point. Decisions can be generated on the basis of this information, and transmitted back to initiate action. At the other extreme, decisions can be taken at the point where the information arises, in accordance with guidelines laid down by a central co-ordinator.

The first model is the one which fits most naturally into the centralizing, Tayloristic, causal mode of thinking. Some human interaction or supervision of the decision-making activity will usually be required: by centralizing it, the interaction can be made at some high level in a hierarchical management structure, where knowledge and authority are assumed to be concentrated, and where the denial of purpose is relaxed. Decisions, also, can be taken on a basis of total information, rather than of partial, localized information.

There are practical difficulties in such a system, which have been at least partly recognized. Gathering and transmitting and processing information centrally can introduce delays which degrade the operation. Human interaction in the centralized decision-making can be hampered by an overload of data, and by the absence of contextual information which is needed to assess the relevance of isolated facts. The chief objection, however, is that if there are other human actors in the system, they become subordinate to the centralized decision-maker interacting with them through the machine.

As seen by the worker, the system gives no information, and allows no control. It issues instructions, the reasons for which are unknown. It embodies Taylor's system under which[5] 'the worker is told minutely just what he is to do and how he is to do it', and it provides this control, moment by moment. This may not have been the intended aim when the system was designed, but the structure of the information system, based upon the causal, Tayloristic view, will ensure that it occurs.

To the worker, the system of interacting machines will constitute one larger machine to which he is subordinate. This is not admissible on the view which has been put forward. We are brought back again to the relation between machines and people: what seems at first sight to be a relation only between machines, implies also a relation between machines and people.

8. *'Assisting' and 'replacing'*

The preceding discussion can be summarized briefly in the following way. On the causal view, machines are designed to replace the human contribution to production. They seldom do so completely, and those workers who are left are required to behave like machines: to stand in place of machines which it has so far proved impossible or uneconomic to produce.

On the purposive view, machines can embody a part of the human purpose of production, but in an imperfect way. Machines are not to be designed to replace the skills and abilities of people, but rather to assist these skills and abilities and make them more productive. The machines should allow existing skills to be relevant, but should not attempt to preserve them in an unchanged form. Scope should be given for skills to change and develop as technology itself develops. Above all, people should never be subservient to machines, but machines should be subservient to people.

These aims, which have been associated with the purposive myth, will no doubt be accepted as desirable by many. Indeed, they are often claimed, at least in part, to be the aims of those who develop new technical systems. For example, 'The expert [computer] system is a high-level intellectual support for the human expert, which explains its other name, *intelligent assistant*.'[6]

As one comes closer to application, however, the emphasis changes:

> This is a revolution, and all revolutions have their casualties. One expert who gladly gave himself and his specialized knowledge over to a knowledge engineer suffered a severe blow to his ego on discovering that the expertise he'd gleaned over the years, and was very well paid and honored for, could be expressed in a few hundred heuristics [computerised 'rules of thumb']. At first he was disbelieving; then he was depressed. Eventually he departed his field, a chastened and moving figure in his bereavement.[7]

This episode reveals no attempt to produce an 'intelligent assistant', working under the direction of the expert. Still less is there an attempt to provide scope for the expert's skill to develop, in such a way that he can perform better than before with the assistance of the computer. 'Assisting the expert' has become 'replacing the expert', and the process by which this change takes place passes unnoticed.

Upon analysis it is easy to see that 'assistance' will always become 'replacement' if we accept the causal myth. The expert's skill is defined to be the application of a set of rules, which express the causal relations determining the expert's behaviour. Assistance then can only mean the application of the same rules by a computer, in order to save the time and effort of the expert. When the rule set is made complete, the expert is no longer needed, because his skill contains nothing more than is embodied in the rules.

It is this destructive metamorphosis of good intentions into ill effects which makes the causal myth so damaging in the context of work. It will be illustrated below in more detail by means of a demonstration project, supported by the ESPRIT programme of the EEC, which had the aim of developing and demonstrating an alternative 'human-centred' technology. The demonstration was successful, but it raises the question whether such a development can resist subversion by the causal myth.

9. *An alternative technology*

The aim of everything that has been said so far has been to support the contention that a different and better technology is possible; one that rejects the Tayloristic assumptions founded upon the causal myth. The reader will naturally ask what this technology would be like, and the question raises a difficulty.

One way of attempting an answer would be to describe a situation many years ahead, when the alternative technology was in full flower. Such efforts can have their own charm, as in William Morris's *News from Nowhere*, but they lack the hard and gritty quality of real life. They are one man's projection of his own ideal future, untempered by the conflicting or co-operating purposes of the others among whom he would have to live and work.

It seems more sensible to admit that we cannot imagine what a remote future may be like. It will arise from the interaction of different interests and different views over the intervening period, and the interests and views will themselves change with time. What we can do with more confidence is to indicate a different direction which we might take from our present situation. This situation is deeply penetrated by Taylorism, and our first attempts will not remove all of its influence. They will not represent an ideal, but a step towards something better. The demonstration project which will be described is of this kind.

The ESPRIT project No. 1217(1199) was the successor to a project supported by the Joint Committee of SERC and ESRC and initiated at UMIST in 1982. This had as its aim the development of a manufacturing system, containing computer-controlled machines, in which 'operators are not subordinate to machines'. The project has been described elsewhere:[8] its practical achievements were severely limited by the resources available, but it served to expose the possibilities, and the problems involved in achieving them.

The ESPRIT project was funded to a level about fifty times greater and involved three countries (Denmark, Germany and the UK) with eleven partners including Universities and industrial companies. Like the UMIST project, it was jointly staffed by social scientists and technologists. The aim was to develop further the ideas underlying the UMIST project, and to carry them to the stage of implementation where they could be demonstrated in industrial conditions. Reports of the work are being prepared and will be published later[9]—only a brief sketch is attempted here.

The title of the ESPRIT project was 'Human-centred computer integrated manufacturing systems', and the term 'human-centred' has become established to describe systems designed in a non-Tayloristic way: systems in which the human contribution and the contribution by machines are considered and worked out together at the design stage. Similar aims in work by social scientists have been described by the term 'sociotechnical design', though social scientists have seldom been offered the opportunity to be deeply involved in the early design decisions, as they were here. What will be described is one of the three demonstration sites of the project, which was set up by BICC in a company situated in the south of England.

The company manufactures high-quality electrical connectors for communication frequencies up to 46 GHz. There are a large number of different standard types of plug and socket. These different types can be associated in many different combinations, in bodies which may, for example, be straight or angled or T-shaped. The number of different possibilities, even after rationalization, is several thousands. Some of these product types are ordered infrequently and in small numbers, and making for stock is then not an economic possibility.

Much of the business of the company therefore consists in responding to a continually varying mix of orders for small and medium quantities. Production in the past had been organized on conventional lines: machines were grouped according to function, with all lathes in one area, all milling machines in another, and so on. Machining work was broken down by the production department into a succession of operations, and a batch of components would be routed (by an accompanying card) from machine to machine as required. The separate parts of a coaxial connector—body, insulator, contact pin—would all be made in this way, and would finally come together for assembly.

At any one time, therefore, there was a multitude of batches of parts in the works, in varying stages of manufacture. Keeping track of the progress of any one order was a matter of great difficulty. Spoiled work at one stage would require urgent changes in scheduling. The delivery time which could be offered to customers was judged to be too long, and performance in meeting these promises was unsatisfactory. At the same time, the amount of work in progress represented an undesirably high capital requirement.

For these reasons, the company felt that it needed to change its methods of working if it was to remain competitive. In this process it also wished to change the working situation, in a way that would involve workers more closely with the aims of the company. It believed that this was the only way to achieve the necessary flexibility of working, and was also more appropriate to the expectations and aspirations of workers.

These intentions can be summarized by saying that the company wished to create a situation in which the purpose of production was shared by the whole workforce, rather than being (as in a Tayloristic system) the exclusive concern of production engineers and management. It was with the aim of achieving this goal that BICC joined the ESPRIT project.

The outcome of the work done during the project was a decision to reorganize production in a small number of 'production islands'. These would (so far as possible) carry out all the machining and assembly tasks on a major component of the connector—on the body, or the insulator and central contact, for example. An island would have all the machines needed for this purpose, would have strong computer support, and would (again so far as possible) have a multiskilled workforce.

Responsibility would be given to the islands for meeting weekly production targets defined centrally, and the way in which this was done would be decided by the island personnel. They would, for example, schedule the sequence in which items were produced during the week to make the best use of machines. They would be responsible for the quality of their production, and would organize the necessary maintenance of machines and the supply of materials. Two islands would have specialized functions which could not be fitted easily into the product oriented islands, namely electroplating and some final assembly.

This reorganization is in progress, but not complete. What will be described is a small prototype cell,* which was an outcome of research during the project, and was operated as a part of normal production in the factory. The layout of the cell is shown in Fig. 9.1.

Within the cell, 100 different kinds of insulator/pin assemblies can be manufactured. Pins, and PTFE insulators, are machined on a numerically controlled Elector-16 lathe, from bar stored in the cell. Programs for the NC lathe can be generated in the cell, using a programmable controller at the lathe. In another part of the ESPRIT project a programmable controller was developed which had facilities especially adapted to a human-centred environment, but this was not available in time for incorporation in the cell. As an alternative to developing them locally, part programs can be generated in a CAD/CAM system and sent to the cell through the computer network.

'Second operations' on pins are carried out on a Combimatic machine, which allows milling, slotting, drilling, etc. It is numerically controlled by a plug board, and is programmed within the cell—computer assistance is given to assist in plugging up the board. Second operations on insulators are done with a manually operated arbor punch.

Assembly of pins and insulators, with epoxy resin where needed, is done within the cell. So also is inspection, using measuring equipment which automatically transmits results to a central computer system. Storage for assemblies, tools, drawings, etc. is provided by a Kardex Industriever vertical carousel. This has a large number of storage trays, arranged to move in a vertical loop under computer control. Stored items or empty trays can be called up from a local computer workstation.

All computer workstations in the cell are linked, and are connected through a factory network to a number of central computers holding factory data and serving CAD and factory planning. A Dextralog station in the cell automatically logs production information and transmits it to a central computer. This information is available to the cell personnel, who can also call up any production information they require about other operations in the factory.

Similarly, inspection data are logged centrally, but can be called up by the cell in numerical or graphical form. A particularly important workstation, placed near the assembly benches, allows scheduling of work within the cell. Work for the coming week is transmitted to this workstation from a

* 'Production island' and 'production cell' are used more or less interchangeably, but with the implication that an island is larger and has responsibility for a larger component, or perhaps a complete product. The island is regarded as a 'factory within a factory'.

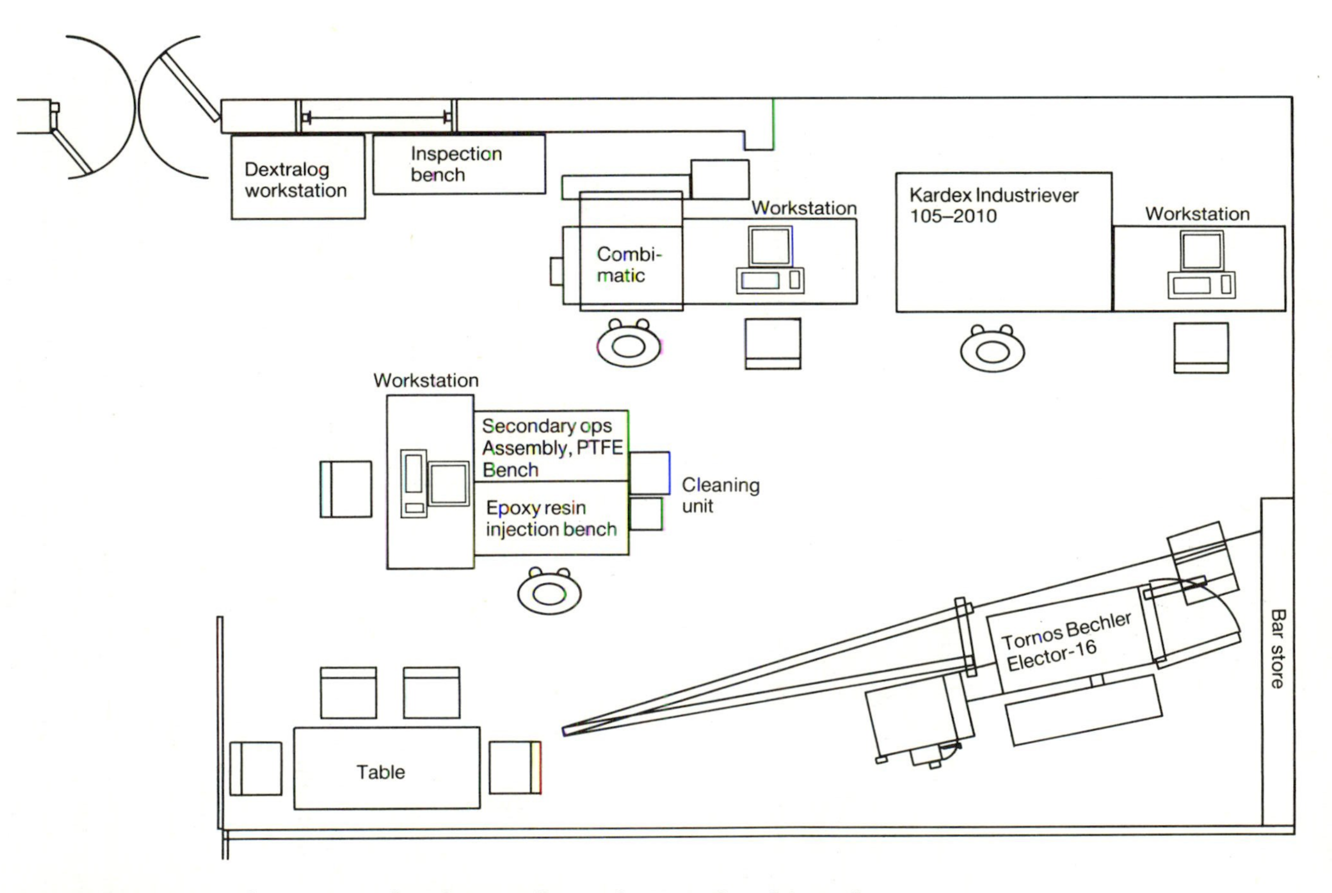

Fig. 9.1 Demonstration human-centred production cell set up by BICC. The cell (a small production island) manufactures pin/insulator assemblies of about 100 types. Computer assistance is given to workers in the cell, but they retain control over local decisions, and responsibility for them.

central computer. Assistance in scheduling the work is given by the workstation, but the final determination is made by the cell personnel.

This describes, very briefly, the physical equipment of the cell and its manner of working. To a large extent, the cell had to be implemented with equipment and software which were designed with a Tayloristic system in mind, and were therefore not ideally suited to a human-centred system. It therefore represents the beginning of a process of change which could lead to significantly different production systems.

The chief factor in the success of such a reorganization is the contribution from the people concerned. Here again there is a troublesome carry-over from previous methods of working. Some managers may be unwilling to see a part of their past responsibility removed and given to the shop floor. Some workers in the islands may be uneasy at accepting wider responsibilities than they have been accustomed to.

There is a transition process to be managed and facilitated, with due attention throughout to the personal desires and problems of those concerned. Recognizing this, the company has sought to produce a cultural change, by developing an interactive education process involving the entire organization.

Notwithstanding the difficulties, it has been possible to show success in economic terms. Island-based production is now fairly widely used in other companies throughout Europe, though without the computer assistance on which many of the human-centred features of the BICC demonstration depend. An analysis by BICC of the extra benefit accruing from these features shows the following gains.

i. On-time delivery increased from 31 per cent to 86 per cent.
ii. Manufacturing lead time approximately halved.
iii. Work-in-progress value reduced by 45 per cent.
iv. Excess working hours required to meet schedules reduced by 75 per cent.

These improvements represent a benefit to the company, and indirectly to society by increasing the wealth which its labour produces. But of course the same thing could be said of the innovations of Taylor and Ford. The critical question is whether this benefit has been achieved at the expense of the working conditions of those concerned, or whether these working conditions have been improved.

The experience of social scientists suggests that a better working situation has indeed been achieved. Workers have more control over their work, the island operates as a social unit, information which is needed about the demands they must meet, and the extent to which they are meeting them, is readily available to those working in the cell. Still more important, a situation has been created where the further evolution of the working situation can be strongly influenced by those who experience it. Having both responsibility and information, they are in the best position to suggest the direction of further change.

In the demonstration system, more has been achieved in changing the relations between person and person than in changing the relation between people and machines. This is not unexpected, because the redesign and development of machines is a slower process than the modification of an organization. Over a longer period, the reorganized production system would begin to demand changes in the machinery for production and communication to match the new working situation. To respond to these demands, a joint social and technical approach to the development and design of advanced technical systems will be required: this as yet is still in the earliest stages of its development.[8]

10. *Can it persist?*

BICC remark that organizations which adhere to conventional methods of work organization see no relevance in the human-centred approach, and that there seems to be no middle ground. Those who are in touch with research work in production engineering will also know that, almost without exception, it is founded upon a Tayloristic view. Moreover, such experiments as have been carried out by social scientists on related ways of organizing work, for example in autonomous groups, have often failed to survive and spread. The experiments have left unchanged the direction in which technology is moving, so that its advance has subverted the organizational changes.

The question therefore arises whether a development such as the one described can persist, and whether it will provide a pattern for the future. The answer which suggests itself, on the basis of the discussion in earlier chapters, is 'not in a society which accepts and acts upon the causal myth'. In such a society it will be subjected to a destructive process of change, in order to bring it into conformity with an established and unquestioned mode of thought.

The critique mounted on the basis of the causal myth would probably begin in the following way. The aim of a company in organizing production is to minimize the cost defined in a wide sense. Material and labour are direct costs, while capital and indirect labour add further costs. But a long delivery date will also impose a cost if it results in loss of business to competitors, and so will poor quality, failure to meet delivery promises, and similar factors.

Minimizing the total cost is therefore a complicated process, which can be done partly by mathematical means, and partly by rules deduced from experience, both of which can be incorporated in a computer. But it is well known that large optimization problems of this kind are difficult to solve, and in certain circumstances it is beneficial to use 'hierarchical optimization'. This splits the problem into a number of smaller optimization problems, each corresponding to a part of the system. Each part is given its own cost formula, which is allocated by a central coordinator. The cost formulae are so chosen by the coordinator that when each part is optimized separately for its own cost formula, the whole system is optimal according to the original criterion.

What the BICC demonstration has disclosed, on this view, is a way of decomposing the large optimization problem for the factory, into a number of smaller problems, one for each production island. Central coordination is very likely to be done by computer, if not now then at some time in the not distant future. But it is proposed that optimization of the way in which the islands do the work allocated to them should be left to the people working in them.

This is unnecessary, and quite undesirable. Unnecessary, because the same computer network which provides information to the personnel in the islands can provide the information for a control computer in each island that optimizes its operation. Undesirable, because workers vary in their ability. Some will achieve less profitable results than others, and the best will fail to do as well as a computer system which incorporates the experience of a group of appropriate experts.

Granted, the critique might continue, not all of this is at present available, and the BICC system may therefore be a useful interim solution. But it would be a mistake on that account to accept it as final, and to relax the search for improvements. These improvements will be obtained by a scientific study of each part of the operation of an island, resulting in the definition of the best way of carrying it out. Once this has been done, the best way so found can be implemented by means of a computer, which will carry it out uniformly and unfailingly.

Research on many aspects of this programme is already under way. Computer-aided design systems at present usually produce an output which needs human assistance in order to generate the program for a numerically controlled machine. It has for a long time been possible to design in a fully three-dimensional way, but this has been complicated and expensive. Future developments will allow it to become the standard method, and no human assistance will then be needed in generating the NC programs, which can be sent direct from the CAD system to the machines. The NC programs generated in this way can call upon research in cutting technology, ensuring that the most economical cutting conditions are used.

At the same time, research is being done into the best ways of generating schedules of work in a computer. Automatic gauging methods will allow continuous monitoring of the dimensions of parts as they are machined, with automatic corrective action. Manual assembly tasks will at some future date become candidates for automation with robots.

The factory of the future will therefore probably retain the island layout used by BICC, and will profit by its advantages in simplifying the control of the whole operation. Those functions in the islands which in the demonstration have been carried out by people will be studied, optimized, computerized and automated. This will perhaps be a slow process, extending far into the future, but the past history of production engineering assures us that it will not stop before it is complete.

In some such way will the critique be mounted. The transition from using machines to assist human work, to using machines to replace all human work, is as inevitable as in an earlier example (Section 8). As before, it stems from the equation 'man is a machine'. If so, we can profitably replace man by a machine, which is totally under our control, amenable to scientific study, and therefore perfectible in a way that man is not.

The purposive myth, on the other hand, would accept that there are many human activities which can with benefit be given to machines. But the human purpose which the machines are intended to fulfil can never in a complete sense be incorporated into them. For this reason it will always be necessary for machines in a production system to be subordinate to people, and to be designed to be so.

Designed in this spirit, machines would incorporate some part of a human purpose. They would take their place in a world where everything was described by its purpose, and where purposiveness was accepted and respected. Machines would aid in fulfilling human purposes, but as subordinate actors carrying out the intentions of human workers. They would assist human skill and allow it to develop into the new skills which the increased competence of machines continually made possible.

Seeing nature as purposive, we should respect the purpose which is expressed naturally in animals and plants and even by inanimate objects, down to the stones fulfilling their purpose expressed by Hamilton's principle. We should have a care that, in subordinating plants and animals and machines to human purposes, we did not so far warp the relation between existing purposes that we damaged the natural world and our own relationship to it.

We might in this way recapture the 'sweet content' that was felt by Izaak Walton. We might also achieve, in producing the goods and services to meet our human needs, a kind of work which enriched our spirit, and which bound us more firmly to nature and to each other. None of this would seem to us to be strange or forced, because it would arise in a natural way as the expression of our scientific understanding of the world. With the outlook of the purposive myth, the BICC demonstration would be a point of departure, a small step in a long and natural progression.

11. *Recapitulation*

This concludes the project upon which we embarked. To avoid misunderstandings, it will be helpful to say what the book has *not* been intended to do.

First, it is not an attack upon science. Rather, it is an attempt to show how the benefits of science can be obtained without those highly undesirable consequences which have so often arisen from it. We need the honesty, and the patience in the face of evidence, which are characteristic of the best science. We need also the coherent intellectual framework, which it provides more strongly than anything previously achieved. We do not, on that account, have to accept with it the preconceptions of the causal myth with their damaging human consequences.

Secondly, it is not a plea for the rejection or restriction of technology. We need the insight which it gives us into man-made systems, not least to undo the damage which past technology has inflicted. It is rather a plea for a different kind of technology which is not antagonistic to people or to the environment. Because technology is so firmly based upon science, the change in technology will have to begin with a changed outlook in science.

Thirdly, it is not intended to propagate a belief that the purposive myth incorporates truths about reality which the causal myth does not. Both myths are descriptions which we impose upon nature, and both carry with them presuppositions for which there is no warrant in the physical evidence. This evidence supports the equivalence class, the theory, to which both myths belong. It does not support any feature in which one myth differs from the other.

From these negatives follow the more positive aspects of our analysis. To be truly consistent, we ought to use the theory, and never any of its myths. But the theory is an abstract and tenuous entity, and it is beyond our intellectual capacity to use it in a practically constructive way. For scientific purposes, we therefore select one representative myth, and use this as though it were the theory. The choice can be made on grounds of convenience and simplicity and fruitfulness, and we shall never be led into scientific error in this way.

The danger is that if we choose the causal myth for its scientific convenience, and forget that it is a myth and not a theory, we are likely to end by acting as though the myth expressed fundamental truths about the world. We shall then see it as natural and right to exploit nature, and exploit the future, and exploit society, to the extent to our ability. Much of what has been written in previous chapters has been intended to establish this connection.

Now it is true that if we are free to choose a representative myth for its scientific convenience, we are equally free to choose a different myth when we consider how we ought to behave, and the ethical problems that this question must raise. The purposive myth has been used in this way in what was written earlier, and it acts as a corrective to the causal myth.

We regard the world as fulfilling a purpose, which explains every observation we can make as completely as the causal myth, being indeed equivalent to it. But a world impregnated with purpose (even if a purpose which we have ourselves constructed to explain our experience) is different in its effect upon us from a causally determined world. Our own purpose finds a natural place among many. We are not given the world to coerce it to our will, but to live in as an organic part of it.

This has been the conclusion to which we have tended, and if the foundations of this different view are not complete, they are well begun. One doubt must remain. Adam Smith remarked that 'the understandings of the greater part of men are necessarily formed by their ordinary employment'.[10] If we continue to use the causal myth for the purposes of science, and therefore also of technology, will it not continue to influence the development of our technology-based society in the same way as before?

This is what Monod foresaw in a future dominated by the causal myth: 'the choice of scientific *practice* . . . has launched the evolution of culture on a one-way path . . . what we see before us is an abyss of darkness.'[11] But if we cannot be sure of the final outcome we have at least a more promising situation than Monod imagined. He saw one unchallenged view of the world, leading inevitably to his abyss. We have suggested a struggle for influence over our behaviour of two views, one leading to the abyss, but the other to a different and more welcoming future.

12. *Non-sexist postscript*

The masculine bias of the English language, arising as it does out of an historical subjection of women, poses difficulties for a book such as this. The subject for the most part is neutral as between the sexes, and if there were a convenient neuter language it would be natural to use it.

Unfortunately the suggested solutions to this difficulty are not happy. 'They' as a neuter singular is clumsy and ambiguous. 'He or she' is not a neuter form: its consistent use when discussing technology would implicate women in what was a male creation, and would necessitate continual asides to define the different roles of men and women in relation to it. That could be valuable, but it is outside the scope of this book. I have therefore followed the traditional practice, recognizing its defects, but seeing nothing better.

At a deeper level, there is a serious question which has not been addressed in the book. I write with a man's experience and outlook, and the ideal of work that I tacitly suggest is based upon traditional forms of men's work, such as that of the blacksmith. Much of the blacksmith's work was routine, but at times, when the red-hot iron was on the anvil, the blacksmith and his striker were totally absorbed. Time stood still and his surroundings receded. These periods of total absorption have been typical of men's activities, from early hunting onwards.

For an obvious reason, women's work in the past has been different. They have cared for young children, which precludes total absorption in work and withdrawal of attention from the world around. Now to suggest on this account that a different kind of work is appropriate to women could be criticized as sexist, and as perpetuating a past discrimination in the allocation of tasks. But equal criticism could be directed against any attempt to impose a man's view of satisfactory work on women, if it is inappropriate to them.

I have accepted fully in what was written earlier that there will be strong individual differences of outlook upon work. I can also accept that there may (but also perhaps may not) be systematic differences of view between men and women. The purposive myth, I believe, could have the same liberating effect upon both, but whether it would lead to the same views on work by women as by men I leave as an open question.

References

1. Charles R. Walker and Robert H. Guest (1952). *The man on the assembly line*, Harvard University Press.
2. David Noble (1983). Present tense technology, parts 1, 2, 3, *Democracy*, Spring, pp. 8–24, Summer, pp. 70–82, Fall, pp. 71–93.
3. D. Michie (editor) (1979). *Expert systems in the microelectronic age*, Edinburgh University Press; Edward A. Feigenbaum and Pamela McCorduck (1984). *The fifth generation*, Pan Books.
4. Henry Ford, in collaboration with Samuel Crowther (1923). *My life and work*, p. 83, Heinemann.
5. F. W. Taylor (1906). *On the art of cutting metals*, p. 55, American Society of Mechanical Engineers.
6. Reference 3, Feigenbaum, p. 86.
7. Reference 3, Feigenbaum, p. 115.
8. H. H. Rosenbrock (editor) (1989). *Designing human-centred technology*, Springer-Verlag; J. Martin Corbett (1989). Technically embedded constraints on job design and how to overcome them, *IFAC Symposium on Skill based automated production*, Vienna, November.
9. ESPRIT 1217(1199) (1989). *Human-centred CIM systems, Deliverable R27*, BICC final report.
10. See Chapter 7, Reference 10.
11. See Chapter 4, Reference 6.

Appendix 1

Technical development

Introduction

This Appendix gives a technical development to support the material given informally in the text. Results are developed as nearly as possible in the same order as they were presented earlier. Reference is made to chapters and sections of the main text, so that a heading numbered 2.3 refers to Section 3 of Chapter 2. Equations are numbered consistently with this notation: for example equations (2.3.1), (2.3.2), etc.

This method of presentation works well enough in general, but does not give the easiest approach to the variational treatment of quantum mechanics. Accordingly, Appendix 2 gives a simple consecutive account of this treatment. It may be more convenient to read Appendix 2 before Section 5.7 of this Appendix.

1.1. *Purpose and causality*

The notions of purpose and causality are differently defined by different authors,[1] because a large part of their meaning arises from the general structure of ideas within which they are introduced. The structure into which we will fit them is that of mathematical systems theory, and some preliminary ideas from that theory are needed before purpose and causality can be defined. They are given here for reference, and are not intended as a condensed account of systems theory.

An *autonomous* system is one which is not subject to influence from outside. A *non-autonomous* (open) system is subject to external influences, variously called inputs, stimuli, exogenous variables, noise, disturbances, etc. according to their nature and the field of application.

A *non-anticipative* system is one in which the past influences future behaviour, but the future does not influence past behaviour. All physical systems are assumed to be non-anticipative.

A *deterministic* system is one in which its past history, together with exact knowledge of any future inputs, implies its future behaviour, exactly and in detail. Newtonian mechanics is deterministic, as Laplace remarked. The recent theory of chaotic systems[2] has shown that the idea of a deterministic system is much less simple than it seems.

The *state* of a system at a given time is a summary of all information about its past history needed to define the influence which this past history has upon the future behaviour of the system. Often a physical system can be described by a set of first-order differential equations. In vector notation these may have the form

$$\dot{x} = \frac{dx}{dt} = f(x,t) \tag{1.1.1}$$

for an autonomous system, or

$$\dot{x} = f(x,t,u) \tag{1.1.2}$$

for a non-autonomous system. Here the n-vector x qualifies as the state, while u represents a vector of inputs.

A *time-reversible* system is one in which reversing the sense of time (i.e. substituting $-t$ for t) gives a new, valid system of the same type. Newtonian physics is time-reversible: if for example we reverse the motions of all bodies in the solar system (which is the effect of reversing the sign of t) we obtain a new, feasible motion of the system, different from its actual motion.

A *stochastic* (probabilistic) system is one in which random, unpredictable variables occur. Stochastic systems are not in general time-reversible: a drop of dye in water diffuses because of the random motion of the liquid, but reversing time gives a kind of behaviour which is never seen. If we take a non-autonomous deterministic system, and subject it to random inputs (disturbances) we obtain a stochastic system.

An *ensemble* of stochastic systems is a very large collection of such systems, which differ only by having different samples of the random variables. They represent an intellectual device for dealing with stochastic systems. We average some quantity in which we are interested over all the members of ensemble. The behaviour of this average, in suitable circumstances, obeys a deterministic equation, and its future can therefore be predicted from a knowledge of its past history (or more simply from a knowledge of its present state, which summarizes that history) together with a knowledge of any future inputs.

A *causal description* of a system is one which defines its future behaviour in terms of its past history (or more simply, of its present state) together with any future inputs. The system is assumed to be non-anticipative, so that a causal description is not ruled out. (We could not have a causal description of a system that was not non-anticipative). The system may be deterministic and autonomous, or deterministic with known inputs: in either case, the equations describing the system are themselves a causal description. If the system is stochastic, we may still be able to have a causal description of the way in which certain ensemble averages behave.

A *purposive description* of a system is a description in terms of an *objective* (goal, purpose) which is to be achieved in the future. In a non-autonomous system (deterministic or stochastic) some of the inputs may be manipulated so that the objective is achieved. This is the orientation of control theory. In an autonomous system (deterministic or stochastic) we can propose an objective and find the form which the equations of the system must have to achieve this objective. If the equations found in this way (the *policy*) describe the actual behaviour of the system, then the proposed objective gives a purposive description of the system.

Broadly speaking, a causal description looks backward, explaining future behaviour in terms of the past. A purposive description looks forward, explaining future behaviour of the system in terms of some objective which it will achieve. Nevertheless, the actual system, when described purposively, remains non-anticipative. The behaviour at each instant depends upon the past (usually summarized by the state). It is the way in which this behaviour depends upon the past (or state) which ensures that the objective will be achieved.

Two comments can be made upon the above definitions. First, they give the substance of the meaning to be attributed to the terms, but this meaning can be refined almost indefinitely: the refinement becomes necessary as different applications are developed. Secondly, the definitions have assumed a mathematical definition of the system, in terms of differential equations or difference equations or the like. A parallel development can be made for systems defined by a sequence of logical decisions. For example, we can develop an optimal policy for car replacement, where an annual decision is made to sell or not to sell.

2.3. *The feedback amplifier*

Let the open-loop gain of the amplifier be a large real number μ, and let the input signal be u volts, while the output is y volts as in Fig. 2.6. Then

$$y=\mu e=\mu\left(u-\frac{y}{100}\right), \tag{2.3.1}$$

whence if the closed-loop gain is g,

$$y=gu=\left(\frac{\mu}{1+\mu/100}\right)u=\left(\frac{\mu}{\mu+100}\right)100u. \tag{2.3.2}$$

Evaluating this for some increasing values of μ we get the following closed-loop gains,

μ	g
10 000	99.01
100 000	99.90
1 000 000	99.99

So when μ becomes large, the closed-loop gain is nearly 100, and changes very little with alterations of μ. This is a simplistic version of the analysis which underlies the remarks on Black's amplifier. It is not intended to suggest that a low-grade amplifier in the forward path, combined with high feedback, is a good formula for the highest quality of music reproduction.

Instability results from delays, which cause the output voltage to lag behind the input. Then at a certain frequency, the output can be delayed by one half-cycle: when the input is at its maximum, the output is at its minimum. Since the error is formed by subtracting some proportion of the output from the input, this means that feedback increases the input to the amplifier, rather than reducing it. For linear time-invariant systems, a quantitative analysis can be given using the Laplace transform.[3] Nonlinear systems present more difficulty,[4] as also do time-varying systems.

2.5 *Dynamic programming*

Bellman's derivation of dynamic programming proceeds in the following way. Let $W(x,t)$ be the minimum amount of fuel used in going from point x at time t to B at t_1. The fuel used in going from x at t to some $x+\delta x$ at $t+\delta t$ is $L(x,v)\delta t$, where L is the rate at which fuel is used when the velocity is v and the position is x. Then

$$W(x,t) = \min_{v} \{L(x,v)\delta t + W(x+\delta x, t+\delta t)\}. \qquad (2.5.1)$$

The quantity within braces on the right-hand side is what has been referred to in the text: the fuel used in going from the present position (here x) to a point D (which here is $x+\delta x$) plus the least amount of fuel (here $W(x+\delta x,t+\delta t)$) needed to go from D at the new time $t+\delta t$ to B at t_1. The minimization with respect to v determines the value of v to be used in the passage from x at t to $x+\delta x$ at $t+\delta t$.

Expanding $W(x+\delta x,t+\delta t)$ by Taylor's theorem, and noting that W is a function of x and t but not of v, (2.5.1) becomes

$$W(x,t) = \min_v \{L(x,v)\delta t + W(x,t) + \frac{\partial W}{\partial x} v\delta t + \frac{\partial W}{\partial t}\delta t\} \qquad (2.5.2)$$

$$0 = \min_v \{\mathrm{L}(x,v) + \frac{\partial W}{\partial x} v + \frac{\partial W}{\partial t}\}. \qquad (2.5.3)$$

This is Bellman's equation.[5]

The development is essentially unchanged if we ask for a maximum rather than a minimum, or if we ask only for a stationary value. In what follows we shall usually ask for a stationary value, replacing 'min' in (2.5.3) by 'stat', but noting that in simple examples the stationary value is often a minimum. We can also allow L to depend upon t, as well as on x and v, with no further difficulty.

Existence and uniqueness pose awkward mathematical problems, but in the simple example considered in the text we can easily see that a solution exists. It will often be unique, though we can certainly invent situations where it is not: for example if A and B are symmetrically placed on opposite sides of a circular island, with still water.

If in (2.5.3) we ask only for a stationary value, rather than a minimum, we can differentiate with respect to v to obtain

$$\frac{\partial L}{\partial v} + \frac{\partial W}{\partial x} = 0. \qquad (2.5.4)$$

If in addition we put, as in the text,

$$L = \frac{1}{2} m v^2 \qquad (2.5.5)$$

where m is the displacement in some suitable units (chosen to avoid a constant of proportionality) then (2.5.4) gives

$$mv + \frac{\partial W}{\partial x} = 0 \qquad (2.5.6)$$

and (2.5.3) gives

$$\frac{1}{2} m v^2 - m v^2 + \frac{\partial W}{\partial t} = 0 \qquad (2.5.7)$$

or

$$\frac{\partial W}{\partial t}-\frac{1}{2m}\left(\frac{\partial W}{\partial x}\right)^2=0 \tag{2.5.8}$$

This is a partial differential equation to determine $W(x,t)$. It should be noted that the factor 1/2 in (2.5.5) is inserted only for mathematical convenience, and has no significant influence on the results. Furthermore, we can always add a constant to W without affecting (2.5.4) and (2.5.8), which contain only derivatives of W.

If we take the origin at A and let β be the coordinate of B, a solution of (2.5.8) is

$$W=\frac{m(\beta-x)^2}{2(t_1-t)} \tag{2.5.9}$$

which gives from (2.5.6)

$$v=\frac{\beta-x}{t_1-t}. \tag{2.5.10}$$

These are what we should expect. The velocity v is the distance to run divided by the time available, and it remains constant. The rate at which fuel is used is given by (2.5.5), and multiplying by the time needed we get the total fuel W given by (2.5.9), which is the OCF.

The analysis has been written out as though x and v were scalars, but it is a simple matter to obtain the vector case. For example, if x,v are vectors with components x_i,v_i, (2.5.3) becomes

$$0=\operatorname*{stat}_{v}\left\{L(x,v)+\sum_j\frac{\partial W}{\partial x_j}v_j+\frac{\partial W}{\partial t}\right\}. \tag{2.5.11}$$

Hence with $L=\frac{1}{2}m\sum_j v_j^2$,

$$\frac{\partial L}{\partial v_i}=mv_i=-\frac{\partial W}{\partial x_i},\ i=1,2,\ldots,n \tag{2.5.12}$$

and

$$\frac{\partial W}{\partial t}-\frac{1}{2m}\sum_j\left(\frac{\partial W}{\partial x_j}\right)^2=0 \tag{2.5.13}$$

Writing β_i for the coordinates of B, when the origin is at A, a solution of (2.5.13) is

$$W=\frac{m}{2(t_1-t)}\sum_j(\beta_j-x_j)^2 \tag{2.5.14}$$

whence (2.5.12) gives for the components of the velocity

$$v_i = \frac{\beta_i - x_i}{t_1 - t} \tag{2.5.15}$$

The contours of the OCF are circles centred on B. Hence the velocity, which by (2.5.12) is normal to these contours, has constant direction. The numerator and denominator of (2.5.15), evaluated at the boat, decrease at the same rate, and the magnitude of its velocity remains constant.

The development has assumed so far that there is no tidal flow. If there is a tide, having velocity components $f_i(x)$ depending on position, then with components v_i of velocity through the water, the total velocity has components $v_i + f_i(x)$. The variables to be chosen to minimize the fuel consumption are again the components v_i of v, so (2.5.11) becomes

$$0 = \underset{v}{\text{stat}} \left\{ L(x,v) + \sum_j \frac{\partial W}{\partial x_j}(v_j + f_j(x)) + \frac{\partial W}{\partial t} \right\} \tag{2.5.16}$$

while we still have $L = \frac{1}{2}m\sum_i v_i^2$. Hence we obtain

$$\frac{\partial L}{\partial v_i} = mv_i = -\frac{\partial W}{\partial x_i} \tag{2.5.17}$$

showing that v_i is proportional to $-\partial W/\partial x_i$. This implies as before that the velocity vector v is in the direction of steepest descent of the function W, which is perpendicular to the local contour of W. The magnitude of v is proportional to the gradient of W. In general, neither the magnitude nor the direction of v will now be constant.

3.2. *Passenger liner*

Let the air conditioning load be represented by $a - gmx_2$, where a and g are constants, m is the displacement of the vessel, and x_2 is the latitude measured from some convenient datum. Then the rate at which fuel is used, for both propulsion and air conditioning, becomes

$$L = \frac{1}{2}mv^2 + a - gmx_2 \tag{3.2.1}$$

and our variational principle is

$$\delta\int_t^{t_1} L(x,v)\mathrm{d}\tau = \delta\int_t^{t_1} (\tfrac{1}{2}mv^2 + a - gmx_2)\mathrm{d}\tau = 0, \tag{3.2.2}$$

where the end-points of the motion are fixed in time and space. Now

$$\int_t^{t_1} a\,\mathrm{d}\tau = a(t_1 - t), \tag{3.2.3}$$

which is unaffected by changes in the way the vessel moves. We can therefore omit the constant a from L, so generalizing a comment made after equation (2.5.8) above. In physical terms, a consumption of fuel for some purpose at a constant, uniform rate does not affect what we have to do to minimize the total fuel consumption.

Equation (2.5.11) holds, with the new expression for L, and we again obtain (2.5.12). But inserting (2.5.12) into (2.5.11) we now obtain

$$\frac{\partial W}{\partial t} - \frac{1}{2m}\sum_j \left(\frac{\partial W}{\partial x_j}\right)^2 - gmx_2 = 0. \tag{3.2.4}$$

A solution of this equation is

$$W = \frac{m}{2(t_1 - t)}(\beta_1 - x_1)^2 - \frac{g^2 m}{6}\left(\frac{1}{2}t_1 - t\right)^3 - gmx_2\left(\frac{1}{2}t_1 - t\right) \tag{3.2.5}$$

and (2.5.12) gives

$$v_1 = -\frac{1}{m}\frac{\partial W}{\partial x_1} = \frac{\beta_1 - x_1}{t_1 - t} \tag{3.2.6}$$

$$v_2 = -\frac{1}{m}\frac{\partial W}{\partial x_2} = g\left(\frac{1}{2}t_1 - t\right). \tag{3.2.7}$$

On integrating (3.2.7), and choosing the constant of integration to make $x_2 = 0$ when $t = 0$, we find

$$x_2 = \frac{1}{2}gt(t_1 - t). \tag{3.2.8}$$

As in (2.5.15), (3.2.6) represents a constant component of velocity in the easterly direction. Equation (3.2.7) represents a component of velocity to the north which is equal to $\frac{1}{2}gt_1$ when $t = 0$ and decreases with constant acceleration $-g$ to become $-\frac{1}{2}gt_1$ when $t = t_1$. The path followed by the ship is a parabola with maximum deviation from the line AB equal to $\frac{1}{8}gt_1^2$, which increases as the total journey time t_1 increases.

3.3. *Motion of a stone*

The notation we have used makes it self-evident that the analysis applies directly to the motion of a stone having mass m. Equation (3.2.1) gives the

Lagrangian, $L=T-V$, if we put $a=0$ and allow g to be the gravitational acceleration. Agreement of the numbers can be obtained by choosing suitable units of measurement in the two cases. Then (3.2.2), again with $a=0$, is Hamilton's principle.

The motion of the stone can now be described in the following way. Hamilton's principle defines a purpose, and the stone moves in the way that is needed to fulfil this purpose. From the purpose we can obtain the policy which will fulfil it, and this defines a set of causal relationships. An alternative way of describing the motion of the stone is to say that it obeys the causal relationships which constitute the policy.

Specifically, from (2.5.12) we have

$$\frac{\mathrm{d}}{\mathrm{d}t}mv_i=\frac{\partial}{\partial t}mv_i+\sum_j v_j\frac{\partial}{\partial x_j}mv_i \tag{3.3.1}$$

$$=-\frac{\partial^2 W}{\partial t\,\partial x_i}+\frac{1}{m}\sum_j\frac{\partial W}{\partial x_j}\frac{\partial^2 W}{\partial x_i\,\partial x_j}. \tag{3.3.2}$$

But (3.2.4) gives

$$\frac{\partial^2 W}{\partial t\,\partial x_i}=\frac{1}{m}\sum_j\frac{\partial W}{\partial x_j}\frac{\partial^2 W}{\partial x_i\,\partial x_j}+gm\delta_{i2}, \tag{3.3.3}$$

where δ_{i2} is the Kronecker delta representing $\partial x_2/\partial x_i$. Then using (3.3.3) in (3.3.2) we get

$$\left.\begin{aligned}\frac{\mathrm{d}}{\mathrm{d}t}mv_1&=0\\ \frac{\mathrm{d}}{\mathrm{d}t}mv_2&=-mg,\end{aligned}\right\} \tag{3.3.4}$$

which are Newton's equations for the motion of the stone: they are causal relations defining the policy which will fulfil the purpose embodied in Hamilton's principle.

We can if we wish introduce the notion of force, and regard (3.3.4) as equations between forces. This is the procedure in the usual development of classical mechanics. Though convenient, and natural in view of our subjective experience of forces, this procedure is not necessary. Instead of visualizing the stone as responding by its motion to impressed forces, we can argue directly from the variational principle, as we did in the text from Hero's principle, and in our analogy of the ship sailing so as to minimize fuel consumption. The stone then, in an unforced manner, moves in the way that makes the 'action' stationary. We can emphasize this point by noting that our ship also obeys equations (3.3.4), but we find no need to regard its motion as the response to a force causing it to accelerate towards the south.

4.3. *The IOW*

For a simple proof that spheres invert into spheres, see Sommerville.[6] For Newton's first law, let a point A move in RW along a straight line with constant velocity. We can represent its position in the complex plane by

$$z = a + vt, \tag{4.3.1}$$

where a and v are complex, and the origin is at the centre of the earth. This inverts to give

$$w = \frac{R^2}{z^*} = \frac{R^2}{a^* + v^* t} \tag{4.3.2}$$

where the star denotes the complex conjugate, and hence

$$\left|\frac{dw}{dt}\right| = \left|\frac{-R^2 v^*}{(a^* + v^* t)^2}\right| = \left|\frac{v}{R^2}\right| |w|^2 \tag{4.3.3}$$

This is the speed with which B moves around its circular path, and is proportional to the square of the distance $|w|$ of B from the centre O of the the earth.

If we lived in IOW, that is if we accepted IOW as giving a 'true representation of the world', we should regard $|w|$ as giving a correct measure of the distance of B from O. In order to reconcile the IOW with the results of measurements, we should also believe that our measuring rods suffered a contraction as they approached O. Then it would be open to an innovative physicist or astronomer in IOW to point out that if we took the bold step of inverting our picture of the world, we could avoid the need for this contraction.

The case is similar for the Fitzgerald–Lorentz contraction. This was introduced to reconcile measurements of the speed of light in our usual picture of the world, in which space and time are entirely distinct. Minkowski, in a geometrical interpretation of special relativity, pointed out that no arbitrary reconciliation of this kind is needed if we imagine ourselves to live in a four-dimensional world, in which transformations can affect both space and time. Four dimensions are needed because the Fitzgerald–Lorentz contraction depends upon velocity, involving time as well as distance.

5.1. *Special relativity*

In special relativity, the motion of a particle can be derived from a variational principle analogous to Hamilton's principle, but with a different formula for L, namely[7]

$$L = -(mc^2 + V)\left(1 - \frac{v^2}{c^2}\right)^{1/2}, \tag{5.1.1}$$

where V is the potential energy, $v^2 = \sum_j v_j^2$, and c is the speed of light, which by a suitable choice of the unit of length can be given the value $c = 1$.

Then the variational principle has the same form as Hamilton's principle,

$$\delta \int_t^{t_1} L(x, v)\mathrm{d}\tau = 0 \tag{5.1.2}$$

with L now given the interpretation (5.1.1) instead of its classical definition $L = T - V$. When v/c is small, we can expand (5.1.1) to give

$$L = \frac{1}{2}mv^2 - V - mc^2 + \frac{1}{2}V\frac{v^2}{c^2} \tag{5.1.3}$$

The first two terms are equal to $T - V$, which constitutes the classical Lagrangian. The term mc^2 is constant, and does not affect the conclusions we draw from (5.1.2), as was noted after equation (3.2.3). The last term in (5.1.3) is negligibly small when v/c is small enough.

In Minkowski's interpretation, $L\,\mathrm{d}t$ is written

$$\begin{aligned} L\,\mathrm{d}t &= -\left(mc + \frac{V}{c}\right)\left[c^2(\mathrm{d}t)^2 - \sum_{j=1}^{3} v_j^2(\mathrm{d}t)^2\right]^{1/2} \\ &= -\left(mc + \frac{V}{c}\right)\left[c^2(\mathrm{d}t)^2 - \sum_{j=1}^{3}\left(\frac{\mathrm{d}x_j}{\mathrm{d}t}\right)^2(\mathrm{d}t)^2\right]^{1/2} \\ &= -\left(mc + \frac{V}{c}\right)\left[c^2(\mathrm{d}t)^2 - \sum_{j=1}^{3}(\mathrm{d}x_j)^2\right]^{1/2}. \end{aligned} \tag{5.1.4}$$

Then a four-dimensional space is introduced with coordinates $q_0 = ct$, $q_i = ix_i$ for $i = 1,2,3$, so that an elementary arc $\mathrm{d}s$ is defined by

$$\mathrm{d}s = \left[\sum_{\nu=0}^{3}(\mathrm{d}q_\nu)^2\right]^{1/2} \tag{5.1.5}$$

$$= \left[c^2(\mathrm{d}t)^2 - \sum_{j=1}^{3}(\mathrm{d}x_j)^2\right]^{1/2}. \tag{5.1.6}$$

With this change of notation, (5.1.2) becomes simply

$$\delta\int -\left(mc+\frac{V}{c}\right)\mathrm{d}s=0, \tag{5.1.7}$$

where the integral is taken between fixed end-points in the four-dimensional space. That is, the quantity $-(mc+V/c)$ is integrated with respect to the arc length s measured along the path of the particle in the space and the resulting integral is stationary. When $V=0$, that is for a particle moving freely, (5.1.7) gives

$$\delta\int \mathrm{d}s=0 \tag{5.1.8}$$

because mc is a constant. The particle therefore takes a path of stationary length through the four-dimensional space.

5.2. *General relativity*

This pleasingly simple geometric result is extended in general relativity to the case where the potential V is obtained from gravitational forces. The formula (5.1.5)

$$(\mathrm{d}s)^2=\sum_{\nu=0}^{3}(\mathrm{d}q_\nu)^2 \tag{5.2.1}$$

holds for a 'Euclidean' space, that is, one having no curvature. For a curved space it generalizes to

$$(\mathrm{d}s)^2=\sum_{\mu,\nu=0}^{3} g_{\mu\nu}\mathrm{d}q_\mu \mathrm{d}q_\nu \tag{5.2.2}$$

with the $g_{\mu\nu}$ functions of the q_ρ.

The symmetric tensor $g_{\mu\nu}$ has ten independent elements, and the way these depend upon the q_ρ is open to choice. Given a suitable choice of the functions $g_{\mu\nu}(q)$ we can propose, when V arises from gravitational forces, to replace (5.1.7) by

$$\delta\int \mathrm{d}s=0. \tag{5.2.3}$$

The particle then, under the influence of gravity, will follow a path of stationary length in the curved, four-dimensional space. The curvature results from the presence of other masses, which influence the motion of the particle by changing the curvature of the space, and so changing the stationary path.

As in the passage from the classical principle of Hamilton to (5.1.7), the choice of the functions $g_{\mu\nu}(q)$ must satisfy strong constraints. In most ordinary circumstances, where (5.1.7) is satisfactory, (5.2.3) must give nearly the same result. But if it is to contribute anything new to physical understanding, (5.2.3) must give different results from (5.1.7) in some significant situations, and these results must be experimentally verifiable.

The choice made by Einstein ensures that (5.2.3) will agree with (5.1.7) for most motions under gravity, but predicts three new phenomena.[8] One is a contribution to the precession of the perihelion of Mercury, which removed a long-standing anomaly. The second concerns the bending of light rays in an intense gravitational field. This was verified by observing the passage of light close to the Sun. The third phenomenon, a red-shift in light emitted by a massive star, was verified in observations of the 'dark companion' of the star Sirius.

The variational principles (5.1.7) and (5.2.3) relate time and position along the path in a four-dimensional space. They have a static appearance, because there is no independent variable to play the role which is played by time in (5.1.2). The discrepancy can be removed by introducing a parameter σ to serve this purpose, so that (5.2.3), for example, becomes

$$\delta\int\left[\sum_{\mu,\nu=0}^{3} g_{\mu\nu}(q)\frac{\mathrm{d}q_\mu}{\mathrm{d}\sigma}\frac{\mathrm{d}q_\nu}{\mathrm{d}\sigma}\right]^{1/2}\mathrm{d}\sigma=0. \tag{5.2.4}$$

Then we can apply dynamic programming and deduce the causal relations which express the policy. These appear as generalizations of Newton's equations, but by themselves they would seem arbitrary and unsatisfactory. It is the respective variational principles which give coherence and simplicity to the results.

5.4. *Stochastic optimal control*

For the sake of a simple and graphic representation, the picture of a fishing fleet given in the text does some violence to the mathematical situation we wish to consider, because boats which are close together will experience nearly the same random motion of the water. We ought to consider an ensemble of systems, in each of which one boat is sailing upon its own copy of the sea with its surrounding land, the copies all being identical except that they have different samples of the random motion. Then two boats in the same geographical position (in their own seas) at the same time will suffer different random motions. Nevertheless, provided the boats remain well separated, we shall expect the picture in the text to represent quite well the system which we now consider.

Let the motion of the boats be governed by an Itô equation

$$\mathrm{d}x = v\,\mathrm{d}t + N\,\mathrm{d}z, \tag{5.4.1}$$

where z is a normalized vector Wiener process,[9] and N is a matrix defined here by the relation

$$NN^{\mathrm{T}} = M = (m_{ij}) = \frac{K}{m}I = \frac{K}{m}(\delta_{ij}). \tag{5.4.2}$$

We ask for the variation of the expected fuel consumption, taken over the fleet, to be stationary:

$$\delta\mathscr{E}\int_t^{t_1} L(x,v)\mathrm{d}\tau = 0. \tag{5.4.3}$$

Then if W is the stationary value of the expected fuel consumption, Bellman's equation becomes[10]

$$\underset{v}{\text{stat}}\left\{\frac{\partial W}{\partial t} + \sum_j \frac{\partial W}{\partial x_j}v_j + \frac{1}{2}\sum_{j,k} m_{jk}\frac{\partial^2 W}{\partial x_j \partial x_k} + L\right\} = 0 \tag{5.4.4}$$

where we take $L = \frac{1}{2}mv^2 - V$. Differentiating for the stationary value, we again obtain (2.5.12),

$$\frac{\partial L}{\partial v_i} = mv_i = -\frac{\partial W}{\partial x_i}, \tag{5.4.5}$$

whence from (5.4.4), using (5.4.2),

$$\frac{\partial W}{\partial t} - \frac{1}{m}\sum_j\left(\frac{\partial W}{\partial x_j}\right)^2 + \frac{K}{2m}\sum_j \frac{\partial^2 W}{\partial x_j^2} + \frac{1}{2m}\sum_j\left(\frac{\partial W}{\partial x_j}\right)^2 - V = 0$$

$$\frac{\partial W}{\partial t} - \frac{1}{2m}\sum_j\left(\frac{\partial W}{\partial x_j}\right)^2 + \frac{K}{2m}\sum_j \frac{\partial^2 W}{\partial x_j^2} - V = 0 \tag{5.4.6}$$

With suitable boundary conditions this equation defines W, which is the OCF.

Now (5.4.6) is nonlinear, but the substitution

$$W = -K \log \psi \tag{5.4.7}$$

gives a linear equation for ψ,

$$K\frac{\partial \psi}{\partial t} + \frac{K^2}{2m}\sum_j \frac{\partial^2 \psi}{\partial x_j^2} + V\psi = 0. \tag{5.4.8}$$

If we specify ψ for all x in the permitted region at some initial time t_0, then this equation determines ψ (and therefore also W) for all such x at all later times.

The probability density P for the boats obeys the equation

$$\frac{\partial P}{\partial t} + \sum_j \frac{\partial P v_j}{\partial x_j} - \frac{K}{2m}\sum_j \frac{\partial^2 P}{\partial x_j^2} = 0, \tag{5.4.9}$$

where v_i is obtained from the solution of (5.4.8) by means of (5.4.5) and (5.4.7). We can again specify P for all x in the permitted region at an initial t_0, and solve (5.4.9) to give P at all such x for all later times.

A specially interesting situation occurs when v, V and P depend upon position, but are independent of time. All boats then follow the same control law, defined by $v(x)$, regardless of the time at which they start. Contrary to what might be expected, this does not mean that the OCF is independent of time. We require

$$v_i = -\frac{1}{m}\frac{\partial W}{\partial x_i} = \frac{K}{m\psi}\frac{\partial \psi}{\partial x_i} = \varphi_i(x), \tag{5.4.10}$$

which integrates to give

$$W = f(x) + g(t) \tag{5.4.11}$$

so that

$$\psi = \exp\{-[f(x) + g(t)]/K\} \tag{5.4.12}$$

where by (5.4.6)

$$\frac{\partial g}{\partial t} = \frac{1}{2m}\sum_j \left[\left(\frac{\partial f}{\partial x_j}\right)^2 - K\frac{\partial^2 f}{\partial x_j^2} \right] + V \tag{5.4.13}$$

In this equation a function of t alone is equated to a function of x alone, so both sides must be equal to a constant, say α, and

$$g(t) = \alpha t + \beta. \tag{5.4.14}$$

Now by substitution in (5.4.9), when $\partial P/\partial t=0$, we find that a solution is

$$P=\gamma e^{-2f(x)/K}=\psi(x,t)\psi(x,-t), \tag{5.4.15}$$

where γ is a normalizing constant and β in (5.4.14) is chosen to satisfy $e^{-2\beta}=\gamma$. It follows that when v, V and P are independent of time, v and P cannot be chosen independently at the initial time. When ψ is given, v is defined by (5.4.10), and P is defined by (5.4.15).

In the text, we have taken $V=0$. We have also referred to 'the number of boats in a region', where a mathematical development would use the probability density P. Provided that the total fleet is large enough to make statistical fluctuations negligible, the number of boats in a region is proportional to P. Finally we notice that since the boats do not influence one another, it makes no difference, when v is independent of t, whether they travel at the same time, or travel at different times.

5.7. *Quantum mechanics*

What has been said about stochastic classical systems prefigures in some respects the properties of quantum-mechanical systems, but there are two important differences. First, the governing equation in elementary quantum mechanics, namely Schrödinger's equation, is complex, whereas the governing equations for classical systems are all real. Secondly, stochastic classical systems have a diffusion term involving the second derivatives, $\partial^2 P/\partial x_i^2$, in the equation (5.4.9) governing the evolution of the probability density. There is no such term in the corresponding quantum-mechanical equation.

Both difficulties can be overcome by a development which will now be described. The simplest starting point is the classical oscillator, for which we can write

$$L=\frac{1}{2}m\dot{x}^2-\frac{1}{2}m\omega^2x^2 \tag{5.7.1}$$

and Hamilton's principle is

$$\delta\int_t^{t_1} L(x,\dot{x})\,d\tau=0, \tag{5.7.2}$$

whence the dynamic programming argument gives

$$\frac{\partial W}{\partial t}-\frac{1}{2m}\left(\frac{\partial W}{\partial x}\right)^2-\frac{1}{2}m\omega^2x^2=0 \tag{5.7.3a}$$

$$\frac{\partial W}{\partial x} = -\frac{\partial L}{\partial \dot{x}} \triangleq -p = -m\dot{x}. \tag{5.7.3b}$$

Solutions of (5.7.3) can be found under various boundary conditions. For example, if $x = x(t)$, $x_1 = x(t_1)$, a solution is

$$W = \frac{m\omega}{2\sin\omega(t_1 - t)}[(x^2 + x_1^2)\cos\omega(t_1 - t) - 2xx_1]. \tag{5.7.4}$$

Solutions of the form $W = f(x) + g(t)$ are readily obtained from (5.7.3) by integrating the equations

$$\left.\begin{aligned} &\frac{\partial W}{\partial x} = m\dot{x} = m\omega\sqrt{(a^2 - x^2)} \\ &\frac{\partial W}{\partial t} \triangleq E = \frac{1}{2}ma^2\omega^2 \end{aligned}\right\} \tag{5.7.5}$$

giving

$$W = -\frac{m\omega a^2}{2}\cos^{-1}\frac{x}{a} + \frac{m\omega x}{2}\sqrt{(a^2 - x^2)} + \frac{ma^2\omega^2 t}{2}, \tag{5.7.6}$$

where constants of integration are omitted because they have no dynamical significance. We also find as in (3.3.4) that (5.7.3) implies

$$\left.\begin{aligned} \ddot{x} + \omega^2 x = 0 \\ x = a\cos(\omega t + \varphi). \end{aligned}\right\} \tag{5.7.7}$$

This last equation is inconvenient because $\dot{x}$ is a double-valued function of x. It is therefore usual to replace x in (5.7.7) by a complex quantity $q = x + \mathrm{i}y$, giving

$$q = ce^{\mathrm{i}\omega t}, \quad \dot{q} = \mathrm{i}\omega q, \quad \ddot{q} + \omega^2 q = 0. \tag{5.7.8}$$

The second of these equations represents a rotation about the origin with angular frequency ω.

We now note that (5.7.8) can be obtained directly by replacing x in (5.7.2) by q. The analysis remains valid after the substitution provided that we ask only for a stationary value of the integral, which is now complex. We thus obtain a variational principle describing the motion of a point (which we call the 'complex image') in the complex plane. If we replace x by q, (5.7.4) and (5.7.6) remain solutions of (5.7.3), but they do not correspond to (5.7.8), which give

$$\frac{\partial W}{\partial t} = E = \frac{1}{2}m\dot{q}^2 + \tfrac{1}{2}m\omega^2 q^2 = 0, \tag{5.7.9}$$

whence, by using $\dot{q} = \mathrm{i}\omega q$ in the second of equations (5.7.3), and integrating,

$$W = -\frac{1}{2}\mathrm{i}\omega m q^2. \tag{5.7.10}$$

The physical variables x, $\dot{x}$ are obtained from q, $\dot{q}$ by taking real parts. The energy of the physical system cannot be obtained in this way, as (5.7.9) demonstrates, because of its non-linear relation to q and $\dot{q}$. These facts are very familiar.

To obtain the quantum-mechanical version of the classical oscillator, we add noise to the complex form of (5.7.2) by writing

$$\delta \mathscr{E} \int_t^{t_1} L(q,v)\, \mathrm{d}\tau = 0 \tag{5.7.11}$$

where $\mathscr{E}$ denotes the expectation and

$$\mathrm{d}q = v\, \mathrm{d}t + (1-\mathrm{i})\left(\frac{\hbar}{2m}\right)^{1/2} \mathrm{d}z \tag{5.7.12}$$

with z a normalized Wiener process and $v = u + \mathrm{i}w$. The notion of a stochastic process in complex variables may be unfamiliar, but z is of course real, while (5.7.12) is simply equivalent to the real equations

$$\left.\begin{aligned} \mathrm{d}x &= u\, \mathrm{d}t + \left(\frac{\hbar}{2m}\right)^{1/2} \mathrm{d}z. \\ \mathrm{d}y &= w\, \mathrm{d}t - \left(\frac{\hbar}{2m}\right)^{1/2} \mathrm{d}z. \end{aligned}\right\} \tag{5.7.13}$$

The change of variables

$$\xi = x + y, \qquad \eta = x - y \tag{5.7.14}$$

gives

$$\left.\begin{aligned} \mathrm{d}\xi &= (u+w)\, \mathrm{d}t \\ \mathrm{d}\eta &= (u-w)\, \mathrm{d}t + \left(\frac{2\hbar}{m}\right)^{1/2} \mathrm{d}z \end{aligned}\right\} \tag{5.7.15}$$

so that the noise is inhomogeneous, affecting only one of the two coordinates introduced in (5.7.14).

The usual dynamic programming argument gives

$$\left.\begin{aligned} \frac{\partial W}{\partial t}-\frac{1}{2m}\left(\frac{\partial W}{\partial q}\right)^2-\frac{i\hbar}{2m}\frac{\partial^2 W}{\partial q^2}-\frac{1}{2}m\omega^2 q^2=0 \\ \frac{\partial W}{\partial q}=-\frac{\partial L}{\partial v}=-mv \end{aligned}\right\} \qquad (5.7.16)$$

where W is now complex. These equations are obtained formally from the one-dimensional version of (5.4.5) and (5.4.6) by substituting q for x, and putting $K=-i\hbar$.

As in the earlier development, the change of variable

$$W=i\hbar\log\psi \qquad (5.7.17)$$

linearizes the first of equations (5.7.16) to give

$$-i\hbar\frac{\partial\psi}{\partial t}+\frac{1}{2m}\left(-i\hbar\frac{\partial}{\partial q}\right)^2\psi+\frac{1}{2}m\omega^2q^2\psi=0. \qquad (5.7.18)$$

This corresponds to (5.4.8), and with one reservation is the form which Schrödinger's equation takes for the quantum oscillator. In (5.7.18) the variable q is complex, whereas in Schrödinger's equation it is replaced by the real variable x. But since ψ, like W, is an analytic function, derivatives with respect to q can be replaced by derivatives with respect to x. Schrödinger's equation is therefore the form which (5.7.18) takes when we restrict attention to the real axis.

A solution of (5.7.18) is[11]

$$\psi=e^{-i\omega t/2}\sum_{n=0}^{\infty}A_nH_n(aq)\exp[-(in\omega t-a^2q^2/2)] \qquad (5.7.19)$$

where A_n are constants, H_n are the Hermite polynomials, and $a^2=m\omega/\hbar$. Corresponding to (5.7.19) we have

$$W=-\frac{1}{2}im\omega q^2+\frac{1}{2}\hbar\omega t+i\hbar\log\left[\sum_{n=0}^{\infty}A_nH_n(aq)e^{-in\omega t}\right], \qquad (5.7.20)$$

which may be compared with (5.7.10).

If only one of the A_n differs from zero, (5.7.20) has the form

$$W = -\frac{1}{2}\mathrm{i}m\omega q^2 + \mathrm{i}\hbar \log A_n H_n(aq) + (n+\tfrac{1}{2})\hbar\omega t$$

$$= f(q) + g(t), \qquad (5.7.21)$$

as in (5.4.11) with x replaced by q. This represents an ensemble with v and P independent of t. We can write

$$H_n(aq) = \gamma \prod_{k=1}^{n} (q - \beta_k), \qquad (5.7.22)$$

where β_k are the n real and distinct zeros of $H_n(\alpha q)$, and then

$$v = -\frac{\mathrm{i}\hbar}{m\psi}\frac{\partial\psi}{\partial q} \qquad (5.7.23)$$

$$= \mathrm{i}\omega q - \frac{\mathrm{i}\hbar}{m}\sum_{k=1}^{n}\frac{1}{q-\beta_k}. \qquad (5.7.24)$$

The ensemble of systems we are considering is a set of complex images, identical except for the fact that each has a different realization of the noise. When $n=0$ in (5.7.24), $v = \mathrm{i}\omega q$ as in (5.7.8). The complex images therefore rotate around the origin with angular frequency ω, and with a superimposed random motion arising from (5.7.13).

So far, we have not considered how the complex images relate to the quantum-mechanical particles which are the object of study. We now postulate that properties of the complex images which are real represent properties of the particles. So the positions of complex images on the real axis (not projections as in the classical case) represent positions of particles. For most purposes in what follows we can therefore restrict our attention to those images which are on the real axis. Any property of these images which is real, such as the probability density, is the corresponding property belonging to the particles.

The equation satisfied by the probability density $P(x,y,t)$ of the images cannot be obtained from (5.4.9) by substituting q for x and $-\mathrm{i}\hbar$ for K, because P is everywhere real, and so not analytic, and therefore $\partial P/\partial q$ does not exist. But directly from (5.7.13)

$$\frac{\partial P}{\partial t} + \frac{\partial Pu}{\partial x} + \frac{\partial Pw}{\partial y} - \frac{\hbar}{4m}\left(\frac{\partial^2 P}{\partial x^2} - 2\frac{\partial^2 P}{\partial x\,\partial y} + \frac{\partial^2 P}{\partial y^2}\right) = 0, \qquad (5.7.25)$$

where the term $\partial P/\partial t$ has been kept for generality, though we know it is zero in our present application. By our postulate, we can obtain information, from observations of the particles, about $P(x,0)$, but not about $P(x,y)$ for

$y \neq 0$. We shall therefore have to obtain $P(x,y)$ for $y \neq 0$ from $P(x,0)$ when this is given.

As (5.7.25) is second-order, we shall need a further condition on the real axis, and we take

$$\left(\frac{\partial^2 P}{\partial x^2}+\frac{\partial^2 P}{\partial y^2}\right)_{y=0}=0 \tag{5.7.26}$$

Then with $\psi=\psi(x,y,t)$ we find (Appendix 2, Section 3.1) that the substitution

$$P(x,0,t)=\psi^*(x,0,t)\,\psi(x,0,t) \tag{5.7.27}$$

reduces (5.7.25) to

$$\left(\frac{\partial P}{\partial t}+\frac{\partial Pu}{\partial x}\right)_{y=0}=0, \tag{5.7.28}$$

which agrees with the standard result for the probability density of the particles. In our present system, derived from (5.7.21), P is independent of time, and $u=0$ on the real axis, so (5.7.28) is satisfied.

Returning to (5.7.23), we see that if the momentum mv of the complex images on the real axis is to be real, and so to represent the momentum of the particles, we must have

$$\left(-\mathrm{i}\hbar\frac{\partial}{\partial x}\right)\psi=a\psi \tag{5.7.29}$$

for some real a. That is, ψ must be an eigenfunction of the Hermitian operator $-\mathrm{i}\hbar\partial/\partial x$ in Hilbert space, while the permitted values of $mv=a$ are the eigenvalues of this operator.

Equation (5.7.23) shows that for the solution (5.7.19), the velocity has an imaginary component on the real axis, and it therefore does not represent the velocity of the particles. On the other hand the average velocity taken over the particles is readily obtained from (5.7.23) and (5.7.27),

$$\bar{v}=\int_{-\infty}^{\infty}Pv\,\mathrm{d}x=\frac{1}{m}\int_{-\infty}^{\infty}\psi^*\psi\left(-\frac{\mathrm{i}\hbar}{\psi}\frac{\partial\psi}{\partial x}\right)\mathrm{d}x \tag{5.7.30}$$

and is real (Appendix 2, Section 3.4). Hence by our postulate, $\bar{v}$ is the average velocity, over all x, of the particles in the ensemble. From this fact it can be shown (Appendix 2, Section 3.5) that the proportion of the particles having each permitted value of v is given, as in the standard theory, by a projection in Hilbert space.

From (5.7.24) we can sketch the flow of complex images, as in Figure 1, which is drawn for $n=2$. Now it can easily be shown (Appendix 2, Section 2) that (5.7.11) and (5.7.12) lead to

$$\mathscr{E}m\,\mathrm{d}v = -\frac{\partial V}{\partial q}\,\mathrm{d}t = -m\omega^2 q\,\mathrm{d}t, \tag{5.7.31}$$

which is valid generally, and in particular for any given n. For $n=0$ the result can be obtained directly from $v=\mathrm{i}\omega q$,

$$\mathscr{E}m\,\mathrm{d}v = m\mathrm{i}\omega\mathscr{E}\mathrm{d}q = m\mathrm{i}\omega v\,\mathrm{d}t = -m\omega^2 q\,\mathrm{d}t. \tag{5.7.32}$$

The terms arising from the sum in (7.7.24) therefore contribute nothing to $\mathscr{E}m\,\mathrm{d}v$.

If, on the other hand, we look at the flow sketched in Figure 1, we are apt to conclude that the complex images are subject to forces emanating from the points $q=\beta_i$. This is because we are accustomed, in a classical system, to equate forces to

$$\frac{\mathrm{d}}{\mathrm{d}t}mv = m\left(\frac{\partial v}{\partial t} + v\frac{\partial v}{\partial q}\right), \tag{5.7.33}$$

which for a term $-\mathrm{i}\hbar/m(q-\beta_k)$ gives

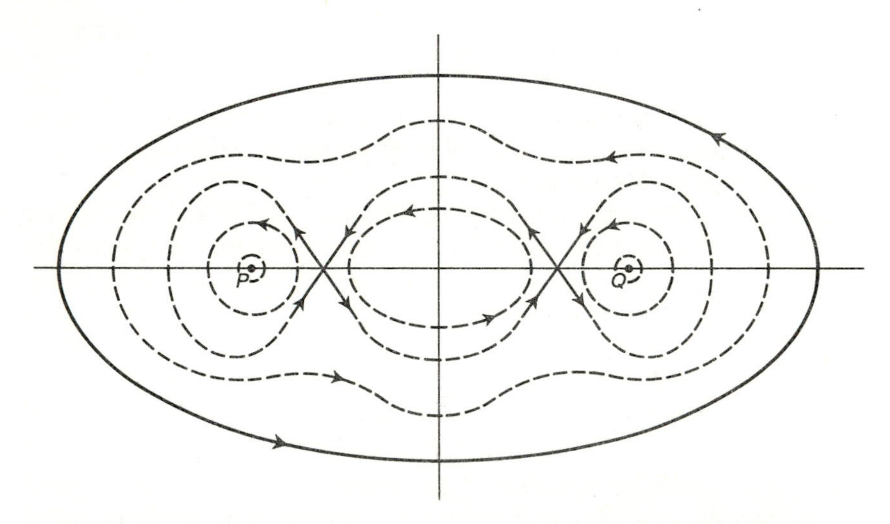

Fig. A1.1 The figure is drawn for $n=2$. From (5.7.24), when $|q|$ is large, v is approximately equal to $\mathrm{i}\omega q$. Near β_k, v is nearly $-\mathrm{i}\hbar/m(q-\beta_k)$. Hence we obtain the portions of the flow shown by solid lines. At some points, P, Q, the velocity on the real axis will be zero. We can complete the picture schematically as shown by the broken lines.

$$\frac{\hbar^2}{m(q-\beta_k)^3} \tag{5.7.34}$$

and when q is real represents a repulsion from the point β_k, inversely proportional to the cube of the distance.

If the particle executing the harmonic motion is an electron in an electrostatic field, there will be one point β_k corresponding to each photon associated with the system. We may then be tempted to associate the 'centres of force' at β_k with the photons, and to say that the repulsive force accounts for the zero probability of finding the electron at these points. But this is, of course, based on the false identification of forces given in (5.7.33). The appropriate expression here, derived from (5.7.12) and leading to (5.7.31), is

$$\mathscr{E} m\,\mathrm{d}v = m\left(\frac{\partial v}{\partial t} + v\frac{\partial v}{\partial q} - \frac{i\hbar}{2m}\frac{\partial^2 v}{\partial q^2}\right)\mathrm{d}t \tag{5.7.35}$$

The term

$$-\frac{i\hbar}{2}\frac{\partial^2 v}{\partial q^2} = \frac{i\hbar}{2m}\frac{\partial^3 W}{\partial q^3} \tag{5.7.36}$$

represents an apparent force which must be added to (5.7.33) to reproduce the motion of the complex images. As was remarked earlier, the notion of force can readily be dispensed with in a variational treatment.

The passage of electrons through holes in a screen is now accounted for in an easy way. What corresponds to our fishing boats in Section 6 of Chapter 5 is not the ensemble of particles but the ensemble of complex images, because it is this which satisfies the variational principle (5.7.11). We have to double the number of dimensions, in effect, by allowing each physical dimension to become complex. This makes visualization impossible, but does not interfere with the mathematical development, from which we obtain W (the OCF), or equivalently the wave function ψ.

The computation by which ψ is obtained is the standard one which is familiar in quantum mechanics. It involves only real values of the coordinates, and ψ is then extended to complex values of the coordinates by analytic continuation. When ψ is available in a closed form, this merely involves replacing x by q. The argument used in Section 6 can be transferred with only verbal changes to the complex images. Then the properties of the particles can be deduced from the postulate given above, and they agree with those obtained in the orthodox manner.

References

1. See for example Mario Bunge (1959). *Causality*, Harvard University Press; Daniel Lerner (editor) (1965). *Cause and effect*, Free Press, New York.
2. Michael Berry (1988). In Lewis Wolpert and Alison Richards (editors), *A passion for science*, pp. 39–51, Oxford University Press; Robert M. May (1984). Simple mathematical models with very complicated dynamics. In Hao Bai-Lin (editor), *Chaos*, pp. 149–57, World Scientific Publishing Company. Note that Figs. 2 and 3 have been interchanged; Joseph Ford (1988). *Chaos: solving the unsolvable, predicting the unpredictable*. In Michael F. Barnsley and Stephen G. Demko (editors), *Chaotic dynamics and fractals*, pp. 1–52, Academic Press.
3. Thomas Kailath (1980). *Linear systems*, Prentice-Hall.
4. For an introductory account, see P. A. Cook (1986). *Nonlinear dynamical systems*, pp. 137–62, Prentice-Hall.
5. Richard E. Bellman and Stuart E. Dreyfus (1962). *Applied dynamic programming*, especially pp. 190–1, Princeton University Press.
6. D. M. Y. Sommerville (1934). *Analytical geometry of three dimensions*, pp. 84–5, Cambridge University Press.
7. Cornelius Lanczos (1970). *The variational principles of mechanics*, pp. 314–30, University of Toronto Press.
8. Reference 7, pp. 330–40.
9. Ludwig Arnold (1974). *Stochastic differential equations*, pp. 35–6, 100–1, Wiley.
10. Reference 9, pp. 211–13.
11. A. Bohm (1979). *Quantum mechanics*, pp. 53–4, Springer-Verlag.

Appendix 2

Brief development of the stochastic variational treatment of quantum mechanics

As a more convenient introduction for the technical reader, this Appendix gives a brief derivation of results in the stochastic variational treatment (SVT) of elementary quantum mechanics.

1. *Schrödinger's equation*

Generalizing Hamilton's principle, we require that

$$\delta\mathscr{E}\int_{t}^{t_1} L(q,v,\tau)\mathrm{d}\tau = 0, \tag{1.1}$$

where $\mathscr{E}$ denotes the expectation, taken over an ensemble of systems each of which can be described by the n-dimensional vector q. Let the Lagrangian be defined by

$$L(q,v,t) = \frac{1}{2}\sum_{j,k} a_{jk}(q,t)v_j v_k + \sum_j a_j(q,t)v_j + a_0(q,t) \tag{1.2}$$

and let

$$\mathrm{d}q = v\,\mathrm{d}t + N(q,t)\mathrm{d}z. \tag{1.3}$$

Here z is a normalized vector Wiener process, and N satisfies

$$NN^{\mathrm{T}} = M = (m_{jk}), \tag{1.4}$$

where M will be defined below. Then Bellman's equation is

$$\underset{v}{\mathrm{stat}}\left\{\frac{\partial W}{\partial t} + \sum_j \frac{\partial W}{\partial q_j}v_j + \frac{1}{2}\sum_{j,k} m_{jk}\frac{\partial^2 W}{\partial q_j \partial q_k} + L\right\} = 0 \tag{1.5}$$

where stat denotes the stationary value and W is the optimal (stationary) value of the expectation of the integral in (1.1).

Differentiating the expression in (1.5) to obtain the stationary value we find

$$\frac{\partial W}{\partial q_i} = -\frac{\partial L}{\partial v_i} \triangleq -p_i \tag{1.6}$$

$$= -\left(\sum_j a_{ij}v_j + a_i\right) \tag{1.7}$$

and if (b_{ij}) is the inverse of the matrix (a_{ij}),

$$v_i = \sum_j b_{ij}(p_j - a_j). \tag{1.8}$$

From (1.5) and (1.6)

$$\frac{\partial W}{\partial t} - \left(\sum_j p_j v_j - L\right) + \frac{1}{2}\sum_{j,k} m_{jk}\frac{\partial^2 W}{\partial q_j \partial q_k} = 0 \tag{1.9}$$

and by use of (1.8) we can eliminate v in favour of p in the expression in parentheses to give

$$\sum_j p_j v_j - L = \sum_{j,k} p_j b_{jk}(p_k - a_k)$$

$$-\left[\frac{1}{2}\sum_{j,k,l,m} a_{jk}b_{jl}(p_l - a_l)b_{km}(p_m - a_m) + \sum_{j,k} a_j b_{jk}(p_k - a_k) + a_0\right]$$

$$= \sum_{j,k} b_{jk}p_j(p_k - a_k) - \frac{1}{2}\sum_{j,k} b_{jk}(p_j - a_j)(p_k - a_k) - \sum_{j,k} b_{jk}a_j(p_k - a_k) - a_0$$

$$= \frac{1}{2}\sum_{j,k} b_{jk}p_j p_k - \sum_{j,k} b_{jk}a_j p_k + \frac{1}{2}\sum_{j,k} b_{jk}a_j a_k - a_0 \triangleq H(q,p,t) \tag{1.10}$$

where we have used $\sum_k a_{jk}b_{km} = \delta_{jm}$, and H is the Hamiltonian.

Now make the transformation

$$W = i\hbar \log \psi \tag{1.11}$$

and set

$$m_{jk} = -i\hbar\frac{\partial^2 H}{\partial p_j \partial p_k} = -i\hbar b_{jk}. \tag{1.12}$$

Then (1.9) becomes

$$\frac{i\hbar}{\psi}\frac{\partial\psi}{\partial t} - \left(\frac{1}{2}\sum_{j,k} b_{jk}p_j p_k - \sum_{j,k} b_{jk}a_j p_k + \frac{1}{2}\sum_{j,k} b_{jk}a_j a_k - a_0\right)$$

$$+\frac{1}{2}\sum_{j,k} b_{jk}\left(p_j p_k + \frac{\hbar^2}{\psi}\frac{\partial^2\psi}{\partial q_j \partial q_k}\right)$$

$$= \frac{i\hbar}{\psi}\frac{\partial\psi}{\partial t} - \left[\frac{1}{2\psi}\sum_{j,k} b_{jk}\left(-i\hbar\frac{\partial}{\partial q_j}\right)\left(-i\hbar\frac{\partial}{\partial q_k}\right)\psi\right.$$

$$\left. - \frac{1}{\psi}\sum_{j,k} b_{jk}a_j\left(-i\hbar\frac{\partial}{\partial q_k}\right)\psi + \frac{1}{2}\sum_{j,k} b_{jk}a_j a_k - a_0\right] = 0. \tag{1.13}$$

Comparison with (1.10) shows that after multiplication by $-\psi$ (1.13) can be written

$$-i\hbar\frac{\partial\psi}{\partial t} + H\left[q_i, \left(-i\hbar\frac{\partial}{\partial q_i}\right), t\right]\psi = 0, \tag{1.14}$$

where we have made a slight change in notation by writing $H(q_i,p_i,t)$ for the Hamiltonian in (1.10).

Together with (1.6) and (1.11), (1.14) defines the motion of an ensemble of points (which we call the complex images) in $\mathbb{C}^n$, subject to (1.1) and (1.3) and to suitable boundary conditions. If $q_i = x_i + iy_i$, and ψ is analytic,

$$\frac{\partial\psi}{\partial q_i} = \frac{\partial\psi}{\partial x_i}. \tag{1.15}$$

Then in the space $\mathscr{S}$ (which is $\mathbb{R}^n$) spanned by the x_i, (1.14) becomes

$$-i\hbar\frac{\partial\psi}{\partial t} + H\left[x_i, \left(-i\hbar\frac{\partial}{\partial x_i}\right), t\right]\psi = 0, \tag{1.16}$$

which is Schrödinger's equation.

If in $\mathscr{S}$ we change from coordinates x_i to ξ_i we have to substitute $\sum_j(\partial\xi_j/\partial x_i)\partial/\partial\xi_j$ for $\partial/\partial x_i$ in (1.16), and it is readily seen that this does not give the same result as we should obtain by starting the analysis from (1.1), (1.3) and (1.12), with L obtained from the classical $L(\xi,\dot{\xi},t)$. The difficulty is inherent in the orthodox treatment of quantum mechanics which we are reproducing, and it is usually dealt with by setting up (1.16) in Cartesian coordinates x_i, and then transforming to the ξ_i. An alternative development has been suggested by Schrödinger.[1] The most desirable formulation would be one that was invariant under Lorentz transformations in Minkowski space.

To avoid confusion, the following notation is now introduced.

The classical momentum is denoted by p.
The SVT momentum defined by (1.6) is denoted by $\tilde{p}$.
The usual quantum-mechanical momentum operator is written $\hat{p}$.

Other physical quantities follow the same rule, so that A, $\tilde{A}$, $\hat{A}$ are the representations of a quantity in classical mechanics, in SVT, and in orthodox quantum mechanics respectively. As W and ψ are the same in SVT and in the orthodox theory, distinguishing marks on them are omitted.

It will be noted that (1.6) and (1.11) give

$$\tilde{p}=\psi^{-1}\hat{p}\psi, \qquad \tilde{p}\psi=\psi\tilde{p}=\hat{p}\psi. \tag{1.17}$$

The Lagrangian in (1.1) has its classical form, so that (1.6) gives

$$\tilde{L}=L(q,\tilde{v},t). \tag{1.18}$$

If we define the Hamiltonian $\tilde{H}$ as the result of eliminating $\tilde{v}$ in favour of $\tilde{p}$ in the expression

$$\sum_j \tilde{p}_j\tilde{v}_j-\tilde{L}, \tag{1.19}$$

then $\tilde{H}$ also has the classical form, with p replaced by $\tilde{p}$. In general, however, the SVT representation of a physical quantity (such as energy, (3.16) below) is not obtained simply by replacing p by $\tilde{p}$ in the classical representation.

2. *Motion of the complex images*

From (1.3) and (1.4) it follows that

$$\mathscr{E}\mathrm{d}\tilde{p}_i=\left(\frac{\partial\tilde{p}_i}{\partial t}+\sum_j\frac{\partial\tilde{p}_i}{\partial q_j}\tilde{v}_j+\frac{1}{2}\sum_{j,k}m_{jk}\frac{\partial^2\tilde{p}_i}{\partial q_j\partial q_k}\right)\mathrm{d}t, \tag{2.1}$$

where by (1.6) and (1.9)

$$\frac{\partial\tilde{p}_i}{\partial t}=-\frac{\partial^2 W}{\partial q_i\partial t}=-\frac{\partial}{\partial q_i}\left(\sum_j\tilde{p}_j\tilde{v}_j-\tilde{L}-\frac{1}{2}\sum_{j,k}m_{jk}\frac{\partial^2 W}{\partial q_j\partial q_k}\right)\mathrm{d}t \tag{2.2}$$

Now eliminate $\tilde{v}$ in favour of $\tilde{p}$ as in (1.10) and regard the resulting $\tilde{H}$ as $\tilde{H}(q,\tilde{p}(q,t),t)$. This gives in (2.2)

$$\frac{\partial \tilde{p}_i}{\partial t} = -\left(\frac{\partial}{\partial q_i}\tilde{H}(q,\tilde{p},t) - \frac{1}{2}\frac{\partial}{\partial q_i}\sum_{j,k} m_{jk}\frac{\partial^2 W}{\partial q_j \partial q_k}\right) \tag{2.3}$$

$$= -\left(\frac{\partial \tilde{H}}{\partial q_i} + \sum_j \frac{\partial \tilde{H}}{\partial \tilde{p}_j}\frac{\partial \tilde{p}_j}{\partial q_i} - \frac{1}{2}\sum_{j,k}\frac{\partial m_{jk}}{\partial q_i}\frac{\partial^2 W}{\partial q_j \partial q_k} - \frac{1}{2}\sum_{j,k} m_{jk}\frac{\partial^3 W}{\partial q_i \partial q_j \partial q_k}\right) \tag{2.4}$$

By Legendre's dual transformation[2]

$$\frac{\partial \tilde{H}}{\partial \tilde{p}_j} = \tilde{v}_j \tag{2.5}$$

while in (2.1)

$$\frac{\partial \tilde{p}_i}{\partial q_j} = -\frac{\partial^2 W}{\partial q_j \partial q_i} = \frac{\partial \tilde{p}_j}{\partial q_i}. \tag{2.6}$$

On using (2.4), (2.5) and (2.6), we obtain from (2.1)

$$\mathscr{E}\mathrm{d}\tilde{p}_i = \left(-\frac{\partial \tilde{H}}{\partial q_i} - \sum_j \tilde{v}_j \frac{\partial \tilde{p}_j}{\partial q_i} + \frac{1}{2}\sum_{j,k}\frac{\partial m_{jk}}{\partial q_i}\frac{\partial^2 W}{\partial q_j \partial q_k} + \frac{1}{2}\sum_{j,k} m_{jk}\frac{\partial^3 W}{\partial q_i \partial q_j \partial q_k}\right.$$

$$\left. + \sum_j \tilde{v}_j \frac{\partial \tilde{p}_j}{\partial q_i} - \frac{1}{2}\sum_{j,k} m_{jk}\frac{\partial^3 W}{\partial q_i \partial q_j \partial q_k}\right)\mathrm{d}t \tag{2.7}$$

$$= -\left(\frac{\partial \tilde{H}}{\partial q_i} + \frac{\mathrm{i}\hbar}{2}\sum_{j,k}\frac{\partial b_{jk}}{\partial q_i}\frac{\partial^2 W}{\partial q_j \partial q_k}\right)\mathrm{d}t. \tag{2.8}$$

If the b_{jk} are all independent of q, (2.5) and (2.8) become

$$\frac{\mathscr{E}\mathrm{d}q_i}{\mathrm{d}t} = \tilde{v}_i = \frac{\partial \tilde{H}}{\partial \tilde{p}_i}, \qquad \frac{\mathscr{E}\mathrm{d}\tilde{p}_i}{dt} = -\frac{\partial \tilde{H}}{\partial q_i} \tag{2.9}$$

which are the analogues of Hamilton's equations. They are outside the framework of orthodox quantum mechanics, where $\tilde{p}$ does not occur, and relate to the motion of the complex images satisfying (1.1) and (1.3), rather than to the physical particles.

3. *Motion of the physical system*

For simplicity, we restrict attention in what follows to an ensemble of systems, each system containing r interacting particles all of mass m. Then

$\tilde{T}$, $\tilde{V}$, W and ψ will be functions on $\mathbb{C}^{3r}$ and summations will run over $3r$ values. In Cartesian coordinates, the matrix (m_{jk}) will become $(-\mathrm{i}\hbar\delta_{jk}/m)$.

Consider the motion of an ensemble of complex images defined by (1.1) and (1.3), and by appropriate initial conditions. We postulate that any physically significant variable which has a real value for the ensemble of complex images has the same value for the ensemble of physical systems (here r particles of mass m). We thus obtain the following results.

3.1 Probability density

In $\mathscr{S}$, the positions of the complex images are real. The probability density of the complex images in $\mathscr{S}$ is therefore the probability density of the particles. From (1.3) we find that the probability density $P(x,y,t)$ of the complex images satisfies

$$0=\frac{\partial P}{\partial t}+\sum_j\frac{\partial P\tilde{u}_j}{\partial x_j}+\sum_j\frac{\partial P\tilde{w}_j}{\partial y_j}-\frac{\hbar}{4m}\sum_j\left(\frac{\partial^2 P}{\partial x_j^2}-2\frac{\partial^2 P}{\partial x_j\partial y_j}+\frac{\partial^2 P}{\partial y_j^2}\right)$$

$$=\left(\frac{\partial P}{\partial t}+\sum_j\frac{\partial P\tilde{u}_j}{\partial x_j}\right)+\sum_j\frac{\partial}{\partial y_j}\left(P\tilde{w}_j+\frac{\hbar}{2m}\frac{\partial P}{\partial x_j}\right)$$

$$-\frac{\hbar}{4m}\sum_j\left(\frac{\partial^2 P}{\partial x_j^2}+\frac{\partial^2 P}{\partial y_j^2}\right) \tag{3.1}$$

where we have written

$$q_j=x_j+\mathrm{i}y_j,\qquad \frac{\tilde{p}_j}{m}=\tilde{v}_j=\tilde{u}_j+\mathrm{i}\tilde{w}_j. \tag{3.2}$$

We also write

$$W=-S+\mathrm{i}\hbar R \tag{3.3}$$

and obtain from (1.6), on noting that W is analytic,

$$\tilde{v}_j=\tilde{u}_j+\mathrm{i}\tilde{w}_j=-\frac{1}{m}\frac{\partial W}{\partial q_j}=-\frac{1}{m}\frac{\partial}{\partial x_j}(-S+\mathrm{i}\hbar R)$$

$$=\frac{\mathrm{i}}{m}\frac{\partial}{\partial y_j}(-S+\mathrm{i}\hbar R) \tag{3.4}$$

whence

$$\left.\begin{aligned} \tilde{u}_j &= -\frac{\hbar}{m}\frac{\partial R}{\partial y_j}, & \tilde{w}_j &= -\frac{\hbar}{m}\frac{\partial R}{\partial x_j} \\ \frac{\partial S}{\partial x_j} &= -\hbar\frac{\partial R}{\partial y_j}, & \frac{\partial S}{\partial y_j} &= \hbar\frac{\partial R}{\partial x_j}. \end{aligned}\right\} \tag{3.5}$$

On using (3.5) in (1.9) and taking the imaginary part, we have in $\mathscr{S}$, where V is real,

$$\frac{\partial R}{\partial t} - \frac{\hbar}{2m}\sum_j \left(\frac{\partial^2 R}{\partial x_j \partial y_j} + 2\frac{\partial R}{\partial x_j}\frac{\partial R}{\partial y_j} \right) = 0. \tag{3.6}$$

Now make the assumption that in $\mathscr{S}$

$$\sum_j \left(\frac{\partial^2 P}{\partial x_j^2} + \frac{\partial^2 P}{\partial y_j^2} \right) = 0, \tag{3.7}$$

and as a tentative solution of (3.1) in $\mathscr{S}$ try

$$P = \psi^* \psi = \exp(-W^*/i\hbar)\exp(W/i\hbar) = e^{2R} \tag{3.8}$$

The first term in parentheses in (3.1) becomes

$$2e^{2R}\left[\frac{\partial R}{\partial t} - \frac{\hbar}{2m}\sum_j \left(\frac{\partial^2 R}{\partial x_j \partial y_j} + 2\frac{\partial R}{\partial x_j}\frac{\partial R}{\partial y_j} \right) \right] \tag{3.9}$$

which is zero in $\mathscr{S}$ by (3.6). The second term in parentheses in (3.1) is

$$-\frac{\hbar}{m}e^{2R}\frac{\partial R}{\partial x_j} + \frac{\hbar}{m}\frac{\partial}{\partial x_j}e^{2R} \tag{3.10}$$

which is zero for all x_j, y_j, so that its derivative with respect to y_j is also zero. From the assumption (3.7) and the results just proved in (3.9) and (3.10), it follows that (3.8) is a solution of (3.1) in $\mathscr{S}$. Moreover, P satisfies the equation in $\mathscr{S}$

$$\frac{\partial P}{\partial t} + \sum_j \frac{\partial P \tilde{u}_j}{\partial x_j} = 0 \tag{3.11}$$

from which it may be computed given its initial condition in $\mathscr{S}$.

Equation (3.11) agrees with the standard result

$$\frac{\partial}{\partial t}\psi^*\psi-\frac{i\hbar}{2m}\sum_j\frac{\partial}{\partial x_j}\left(\psi^*\frac{\partial\psi}{\partial x_j}-\psi\frac{\partial\psi^*}{\partial x_j}\right)=0 \tag{3.12}$$

which can be written

$$\frac{\partial P}{\partial t}+\frac{1}{2}\sum_j\frac{\partial}{\partial x_j}P\left(-\frac{i\hbar}{m\psi}\frac{\partial\psi}{\partial x_j}+\frac{i\hbar}{m\psi^*}\frac{\partial\psi^*}{\partial x_j}\right)$$

$$=\frac{\partial P}{\partial t}+\frac{1}{2}\sum_j\frac{\partial}{\partial x_j}P(\tilde{v}_j+\tilde{v}_j^*)$$

$$=\frac{\partial P}{\partial t}+\sum_j\frac{\partial P\tilde{u}_j}{\partial x_j}=0. \tag{3.13}$$

What has been demonstrated here is that subject to (3.7), $\psi^*\psi$ satisfies the same equation (3.1) in $\mathscr{S}$ as P. Then by (3.13), if P and $\psi^*\psi$ are equal everywhere in $\mathscr{S}$ for some t, they will be equal everywhere in $\mathscr{S}$ for all t. Adopting the spirit of the orthodox theory, we could regard (3.7) as a postulate (or axiom). In the spirit of SVT, it would be more satisfactory if it could be derived from (1.1) and (1.3).

It will be noticed that if we substitute $P=\psi^*\psi=e^{2R}$ in (3.7), the quantity within parentheses can be evaluated, and in general it is not zero. This is to be expected. We know from the orthodox theory that P and $\psi^*\psi$ coincide in $\mathscr{S}$, but we expect them to differ elsewhere, because $\psi^*\psi$ must be infinite somewhere in $\mathbb{C}^{3r}$. Then $\partial^2P/\partial y_j^2$ will differ from $\partial^2\psi^*\psi/\partial y_j^2$ in $\mathscr{S}$, even though $P=\psi^*\psi$ in $\mathscr{S}$.

3.2 Quantization

Suppose that at some time t, all the systems of complex images in the ensemble have the same real value a_i for $\tilde{p}_i$ in $\mathscr{S}$. That is, by (1.6) and (1.11),

$$-\frac{i\hbar}{\psi}\frac{\partial\psi}{\partial x_i}=a_i \tag{3.14}$$

or

$$\left(-i\hbar\frac{\partial}{\partial x_i}\right)\psi\triangleq\hat{p}_i\psi=a_i\psi. \tag{3.15}$$

It follows that all the corresponding particles have momentum $\tilde{p}_i = a_i$, and a_i is an eigenvalue of the operator in (3.15), while ψ is a corresponding eigenfunction.

Similarly, if all systems of complex images have energy $\tilde{E} = a$ in $\mathscr{S}$, we have in terms of W and the initial time t

$$\tilde{E} \triangleq \frac{\partial W}{\partial t} = \frac{i\hbar}{\psi}\frac{\partial \psi}{\partial t} = V - \frac{\hbar^2}{2m\psi}\sum_j \frac{\partial^2 \psi}{\partial x_j^2} = a \tag{3.16}$$

by (1.9) and (1.11). Hence

$$\left[V + \frac{1}{2m}\sum_j \left(-i\hbar\frac{\partial}{\partial x_j}\right)^2\right]\psi = \left[V + \frac{1}{2m}\sum_j \hat{p}_j^2\right]\psi \triangleq \hat{E}\psi = a\psi \tag{3.17}$$

so that all the physical systems have energy a, and a is an eigenvalue of the operator in (3.17), while ψ is a corresponding eigenfunction. We are here neglecting the possibility of discrepancies such as the one defined by eqn. (5.7.9) of Appendix 1.

Equations such as (3.15) and (3.17) define certain values of a, the eigenvalues of the operators, which are allowable values for the corresponding physical variables. The permissible values of a may be continuous or discrete, and in the second case they imply quantization.

3.3 Interference

Let particles (with position denoted by a three-vector x) arrive, from a source on the left, at a screen in which there are two non-overlapping holes, A and B. With hole A alone open, there will be a known $\psi_A(x,t)$ at the hole, obtained from a solution of Schrödinger's equation for the region to the left. Elsewhere at the screen, ψ will be zero by (3.8) above. With this boundary condition, $\psi_A(x,t)$ can be extended to the whole region to the right of the screen by Schrödinger's equation. Hence (though we do not need this result) $\psi_A(x,t)$ can be extended by analytic continuation to complex values of $q = x + iy$.

Similarly, when hole B alone is open, there will be some $\psi_B(x,t)$ for the region to the right of the screen. When both holes are open, there will be some $\psi(x,t)$ for the region to the right of the screen, which will be a solution of Schrödinger's equation satisfying the boundary conditions at the screen. Taking account of normalization, the appropriate ψ is

$$\psi(x,t) = c_A\psi_A(x,t) + c_B\psi_B(x,t) \tag{3.18}$$

This follows from two facts:

(i) Schrödinger's equation is linear, so the weighted sum of the solutions ψ_A and ψ_B is also a solution.

(ii) At the screen, $P_A = \psi_A^* \psi_A$ is zero everywhere except in hole A, and $P_B = \psi_B^* \psi_B$ is zero everywhere except in hole B. Hence at the screen, one or other of ψ_A and ψ_B is always zero. It follows that the probability density at the screen is

$$\begin{aligned} P = \psi^* \psi &= (c_A^* \psi_A^* + c_B^* \psi_B^*)(c_A \psi_A + c_B \psi_B) \\ &= c_A^* c_A P_A + c_B^* c_B P_B = \gamma_A P_A + \gamma_B P_B, \end{aligned} \tag{3.19}$$

which is what we require. Equation (3.19) will not be true in the region to the right of the screen, where ψ_A and ψ_B will both in general be non-zero.

This analysis applies when there is no way of identifying which hole a particle has passed through, and when a single ψ has to be used for all particles to the right of the screen. When the passage of a particle can be assigned to a particular hole, the different functions ψ_A and ψ_B can be used according to which hole has served. Then, as in the classical case, the existence of hole B (resp. A) does not affect the motion of particles which pass through hole A (resp. B). Everywhere to the right of the screen P will be given by $\gamma_A P_A + \gamma_B P_B$, with appropriate γ_A and γ_B, but ψ defined by (3.18) will have no significance.

3.4 Mean values

In Section (3.3) above, let x_1 be normal to the screen, and in the region to the right of the screen let

$$\left.\begin{aligned} \tilde{p}_1^A &= -\frac{i\hbar}{\psi_A}\frac{\partial \psi_A}{\partial x_1} = a_A \\ \tilde{p}_1^B &= -\frac{i\hbar}{\psi_B}\frac{\partial \psi_B}{\partial x_1} = a_B, \end{aligned}\right\} \tag{3.20}$$

where a_A and a_B are distinct real constants, and the indices A and B have been raised on vectors for convenience. Then corresponding to (3.18) we have

$$\begin{aligned} \tilde{p}_1 &= -\frac{i\hbar}{\psi}\frac{\partial \psi}{\partial x_1} \\ &= \frac{c_A \psi_A a_A + c_B \psi_B a_B}{c_A \psi_A + c_B \psi_B}. \end{aligned} \tag{3.21}$$

This $\tilde{p}_1$ cannot be a real constant, because then $c_A \psi_A + c_B \psi_B$ would be an eigenfunction of the operator $-i\hbar\partial/\partial x_1$, and this cannot be so because ψ_A and ψ_B are eigenfunctions corresponding to distinct eigenvalues a_A, a_B. Therefore the momentum $\tilde{p}_1$ of the complex images in $\mathscr{S}$ given by (3.21) is not the momentum of the particles.

On the other hand, the mean momentum of the images in $\mathscr{S}$ in the direction of x_1 is

$$\bar{\tilde{p}}_1 = \int P(x)\tilde{p}_1(x)\mathrm{d}^3x$$
$$= \int (c_A^*\psi_A^* + c_B^*\psi_B^*)(c_A\psi_A a_A + c_B\psi_B a_B)\mathrm{d}^3x, \quad (3.22)$$

where the integrals extend over all x. Now ψ_A and ψ_B are eigenfunctions of the Hermitian operator $-i\hbar\partial/\partial x_1$ corresponding to distinct eigenvalues, and therefore are orthogonal, so that (3.22) gives

$$\bar{\tilde{p}}_1 = c_A^*c_A a_A + c_B^*c_B a_B \quad (3.23)$$

This mean momentum of the complex images is real, and is therefore the mean momentum of the particles.

3.5 Measurement

Define a process of measuring the momentum of particles in the direction of x_1 which is the converse of the process in Section (3.3) of combining two sub-ensembles. That is, suppose that we can in some way decompose a given ensemble of particles into sub-ensembles, all particles belonging to each sub-ensemble having the same momentum in the x_1-direction. Suppose also that all the members of each sub-ensemble, and only those, are made to pass through one hole in a screen which we provide, there being one such (non-overlapping) hole for each sub-ensemble.

Then the computation by which ψ_A and ψ_B were extended to the right of the screen in Section (3.3) can be applied in reverse. An ensemble having the ψ given in (3.18), and satisfying (3.20), will give rise to two sub-ensembles. In one sub-ensemble, all particles will have $\tilde{p}_1 = a_A$ and in the other they will have $\tilde{p}_1 = a_B$. Since the mean momentum taken over all particles is given by (3.23), it follows that the particles in the two sub-ensembles will be in the proportions $c_A^*c_A$ and $c_B^*c_B$.

This scheme can be generalized to other measured variables and to other schemes of measurement. Given a general ψ we proceed as follows. Find a complete orthonormal set of eigenfunctions u_i of the operator corresponding to the measured variable. The corresponding eigenvalues comprise all values of the measured variable which can be taken by the particles. Express ψ in the form

$$\psi = \sum_j c_j u_j \quad (3.24)$$

for discrete eigenvalues, and in the corresponding integral form for continuous eigenvalues. Then the probability of occurrence of the measured value a_i is $c_i^* c_i$, due account being taken of multiple eigenvalues. This is the standard procedure.

4. *Poisson brackets and the commutator*

Let two physical variables in SVT be defined by

$$\tilde{A} = a_0(q) + \sum_j a_j(q)\tilde{p}_j, \qquad \tilde{B} = b_0(q) + \sum_j b_j(q)\tilde{p}_j \tag{4.1}$$

The Poisson bracket for $\tilde{A}$, $\tilde{B}$ is

$$\{\tilde{A}, \tilde{B}\} = \sum_j \left(\frac{\partial \tilde{A}}{\partial q_j} \frac{\partial \tilde{B}}{\partial \tilde{p}_j} - \frac{\partial \tilde{B}}{\partial q_j} \frac{\partial \tilde{A}}{\partial \tilde{p}_j} \right) \tag{4.2}$$

and by a straightforward but tedious calculation we can show that

$$\{\tilde{A}, \tilde{B}\}\psi = \frac{1}{\mathrm{i}\hbar}[\hat{A}, \hat{B}]\psi \tag{4.3}$$

where $\hat{A}, \hat{B}$ are obtained from (4.1) by replacing $\tilde{p}_j$ by the operator $\hat{p}_j = -\mathrm{i}\hbar\partial/\partial q_j$.

As an example, for angular momenta we have

$$\tilde{m}_3\psi = (q_1\tilde{p}_2 - q_2\tilde{p}_1)\psi = \{q_2\tilde{p}_3 - q_3\tilde{p}_2, q_3\tilde{p}_1 - q_1\tilde{p}_3\}\psi$$

$$= \{\tilde{m}_1, \tilde{m}_2\}\psi = \frac{1}{\mathrm{i}\hbar}[\hat{m}_1, \hat{m}_2]\psi$$

$$= (q_1\hat{p}_2 - q_2\hat{p}_1)\psi = \hat{m}_3\psi \tag{4.4}$$

and similarly for $\tilde{m}_1$ and $\tilde{m}_2$. Equation (4.4) can be verified directly from the second and sixth items in the chain of equalities.

The result in (4.3) gives a justification within SVT for the standard replacement of the Poisson bracket by the commutator, valid for the particular variables defined in (4.1). The result can be generalized, but not to the extent that might be supposed. For example, from

$$\tilde{A} = a_0(q), \qquad \tilde{B} = \frac{1}{2}\tilde{p}^2 \tag{4.5}$$

we obtain

$$\{a_0, \frac{1}{2}\tilde{p}^2\}\psi = -\mathrm{i}\hbar \frac{\partial a_0}{\partial q}\frac{\partial \psi}{\partial q} \tag{4.6}$$

while

$$\frac{1}{\mathrm{i}\hbar}[a_0, \frac{1}{2}\hat{p}^2]\psi = -\mathrm{i}\hbar \frac{\partial a_0}{\partial q}\frac{\partial \psi}{\partial q} - \frac{\mathrm{i}\hbar}{2}\frac{\partial^2 a_0}{\partial q^2}\psi \tag{4.7}$$

The results in (4.6) and (4.7) agree when $a_0(q) = q$, extending (4.3) to a case not considered before, but they disagree when $a_0(q) = q^2$. An explanation of the discrepancy will be found in the following section.

5. *Time derivative*

As in (4.1) we let

$$\tilde{A} = a_0(q) + \sum_j a_j(q)\tilde{p}_j \tag{5.1}$$

and we take

$$\tilde{H} = \frac{1}{2m}\sum_j \tilde{p}_j^2 + V. \tag{5.2}$$

We also note that by previous results

$$\left.\begin{aligned} m\tilde{v}_i = \tilde{p}_i = -\frac{\partial W}{\partial q_i} = \psi^{-1}\left(-\mathrm{i}\hbar\frac{\partial}{\partial q_i}\right)\psi = \psi^{-1}\hat{p}_i\psi \\ \frac{\partial}{\partial q_i} = \frac{\mathrm{i}}{\hbar}\hat{p}_i. \end{aligned}\right\} \tag{5.3}$$

Then treating $\tilde{A}(q,\tilde{p}) = \tilde{A}[q, -\partial W(q,t)/\partial q]$ as a function of q and t,

$$\frac{\mathscr{E}\mathrm{d}\tilde{A}}{\mathrm{d}t} = \frac{\partial \tilde{A}}{\partial t} + \sum_j \frac{\partial \tilde{A}}{\partial q_j}\frac{\tilde{p}_j}{m} - \frac{\mathrm{i}\hbar}{2m}\sum_j \frac{\partial^2 A}{\partial q_j^2} \tag{5.4}$$

On using (5.1) and (5.2), together with (1.9), we obtain from (5.4) after some manipulation

$$\left(\frac{\mathscr{E}\mathrm{d}\tilde{A}}{\mathrm{d}t}\right)\psi = \frac{1}{\mathrm{i}\hbar}\left\{\sum_j \left[a_j(\hat{p}_j V)\psi - \frac{1}{m}(\hat{p}_j a_0)(\hat{p}_j\psi) - \frac{1}{2m}(\hat{p}_j^2 a_0)\psi\right]\right.$$

$$-\frac{1}{2m}\sum_{j,k}\left[(\hat{p}_j^2 a_k)(\hat{p}_k\psi)+2(\hat{p}_j a_k)(\hat{p}_j\hat{p}_k\psi)\right]\Bigg\} \tag{5.5}$$

If on the other hand we write

$$\frac{1}{i\hbar}[\hat{A},\hat{H}]\psi=\frac{1}{i\hbar}\left\{\left(a_0+\sum_j q_j\hat{p}_j\right)\left(\frac{1}{2m}\sum_j\hat{p}_j^2+V\right)\psi\right.$$

$$\left.-\left(\frac{1}{2m}\sum_j\hat{p}_j^2+V\right)\left(a_0+\sum_j a_j\hat{p}_j\right)\psi\right\} \tag{5.6}$$

and evaluate this expression, we find the same result as on the right-hand side of (5.5). Hence with the assumed $\tilde{A}$ and $\tilde{H}$, and with the definition (5.6),

$$\left(\frac{\mathscr{E}\mathrm{d}\tilde{A}}{\mathrm{d}t}\right)\psi=\frac{1}{i\hbar}[\hat{A},\hat{H}]\psi \tag{5.7}$$

which corresponds to the standard result.

On the other hand, if we let $\tilde{A}$ in (5.4) be any function of q and $\tilde{p}$, we have

$$\frac{\mathscr{E}\mathrm{d}\tilde{A}}{\mathrm{d}t}=\sum_j\frac{\partial\tilde{A}}{\partial q_j}\frac{\mathscr{E}\mathrm{d}q_j}{dt}+\sum_j\frac{\partial\tilde{A}}{\partial\tilde{p}_j}\frac{\mathscr{E}\mathrm{d}\tilde{p}_j}{\mathrm{d}t}-\frac{i\hbar}{2m}\sum_j\left\{\frac{\partial^2\tilde{A}}{\partial q_j^2}+2\sum_k\frac{\partial^2\tilde{A}}{\partial q_j\partial\tilde{p}_k}\left(\frac{\partial\tilde{p}_k}{\partial q_j}\right)\right.$$

$$\left.+\sum_{k,l}\frac{\partial^2\tilde{A}}{\partial\tilde{p}_k\partial\tilde{p}_l}\left(\frac{\partial\tilde{p}_k}{\partial q_j}\right)\left(\frac{\partial\tilde{p}_l}{\partial q_j}\right)\right\} \tag{5.8}$$

$$=\{\tilde{A},\tilde{H}\}-\frac{i\hbar}{2m}\sum_j\{\ \} \tag{5.9}$$

by (2.9). Then (5.7) gives for the particular $\tilde{A}$ in (5.1)

$$\left(\frac{\mathscr{E}\mathrm{d}\tilde{A}}{\mathrm{d}t}\right)\psi=\frac{1}{i\hbar}[\hat{A},\hat{H}]\psi=\{\tilde{A},\tilde{H}\}\psi$$

$$-\frac{i\hbar\psi}{2m}\sum_j\left\{\frac{\partial^2 a_0}{\partial q_j^2}+\sum_k\frac{\partial^2 a_k}{\partial q_j^2}\tilde{p}_k+2\sum_k\frac{\partial a_k}{\partial q_j}\frac{\partial\tilde{p}_k}{\partial q_j}\right\} \tag{5.10}$$

where $\tilde{p}_k=(-i\hbar/\psi)\partial\psi/\partial q_k$; and the Poisson bracket can be substituted for the commutator only when the last term in (5.10) is zero. Equations (4.6) and (4.7) illustrate (5.10) for the simple case where $\tilde{A}=a_0(q)$, and $\tilde{B}$ is the SVT Hamiltonian for a unit mass subject to no forces.

The result in (5.7) can be considerably extended. We find, for example, that when (5.2) holds,

$$\psi\mathscr{E}\mathrm{d}[b_k(q)\psi^{-1}\hat{p}^k\psi]=\frac{1}{\mathrm{i}\hbar}[b_k(q)\hat{p}^k,\hat{H}]\psi\ \mathrm{d}t. \tag{5.11}$$

Hence, if $\hat{B}(q,\hat{p})$ can be expanded as

$$\hat{B}(q,\hat{p})\psi=\sum_k b_k\hat{p}^k\psi=\psi\sum_k b_k\psi^{-1}\hat{p}^k\psi=\psi\tilde{B} \tag{5.12}$$

we have

$$\left(\frac{\mathscr{E}\mathrm{d}\tilde{B}}{\mathrm{d}t}\right)\psi=\frac{1}{\mathrm{i}\hbar}[\hat{B},\hat{H}]\psi \tag{5.13}$$

Here we have written $\tilde{B}=\sum_k b_k\psi^{-1}\hat{p}^k\psi$, noting that when k can exceed 1, $\tilde{B}$ is not a function of $\tilde{p}$, but is formed by analogy with (3.16).

Within SVT there are therefore two different justifications which can be given, in two different circumstances, for replacing the Poisson bracket by the commutator. In some circumstances, the Poisson bracket gives the commutator directly, as in (4.3). In certain other circumstances, illustrated by (4.6) and (4.7), this does not hold; but the commutator takes over the role of the Poisson bracket by representing a rate of change, as in (5.7) and (5.13).

6. *Mixed systems*

In the preceding development, the variables were all quantum-mechanical. By an appropriate modification it is possible to consider systems in which some variables are quantum-mechanical, and some are treated as classical. The only point requiring care is that ψ in (1.11) becomes undefined when $\hbar=0$.

Suppose for example that we have a system for which the classical Hamiltonian is

$$H=\frac{1}{2}\sum_{j,k=1}^{2} b_{jk}p_jp_k+V(q). \tag{6.1}$$

We wish to regard p_1 as classical, but to let p_2 become quantum-mechanical. Then in determining p_1, we neglect the effect upon it of the quantum-mechanical part of the system, and use Hamilton's principle to obtain

$$\frac{\partial W_c}{\partial t}=\frac{1}{2}b_{11}p_1^2+V \tag{6.2}$$

where $p_1=-\partial W_c/\partial q_1$.

Treating the whole system as quantum-mechanical, we should obtain from (1.9)

$$\frac{\partial W_{qm}}{\partial t}=\frac{i\hbar}{\psi}\frac{\partial\psi}{\partial t}=\frac{1}{2}(b_{11}\tilde{p}_1^2+b_{12}\tilde{p}_1\tilde{p}_2)-\frac{i\hbar}{2}\left(b_{11}\frac{\partial\tilde{p}_1}{\partial q_1}+b_{12}\frac{\partial\tilde{p}_2}{\partial q_1}\right)$$

$$+\frac{1}{2}(b_{12}\tilde{p}_1\tilde{p}_2+b_{22}\tilde{p}_2^2)-\frac{i\hbar}{2}\left(b_{12}\frac{\partial\tilde{p}_1}{\partial q_2}+b_{22}\frac{\partial\tilde{p}_2}{\partial q_2}\right)+V \tag{6.3}$$

where we have used $b_{12}=b_{21}$. Now

$$-\frac{i\hbar}{2}\left(b_{11}\frac{\partial\tilde{p}_1}{\partial q_1}+b_{12}\frac{\partial\tilde{p}_2}{\partial q_1}\right)$$

$$=-\frac{1}{2}(b_{11}\tilde{p}_1^2+b_{12}\tilde{p}_1\tilde{p}_2-b_{11}\psi^{-1}\hat{p}_1^2\psi-b_{12}\psi^{-1}\hat{p}_1\hat{p}_2\psi). \tag{6.4}$$

so that its effect is to substitute $\psi^{-1}\hat{p}_j\hat{p}_k\psi$ for $\tilde{p}_j\tilde{p}_k$ in (6.3), and similarly for the second line in that equation. It is this substitution, together with (1.17), which converts the classical to the quantum-mechanical description, and we wish to apply it selectively so as to replace $\tilde{p}_2$ while retaining $\tilde{p}_1$. The latter is put equal to p_1 determined from (6.2).

To achieve this result, we delete the term in (6.3) which occurs on the left-hand side of (6.4), leaving

$$\frac{i\hbar}{\psi}\frac{\partial\psi}{\partial t}=\frac{1}{2}(b_{11}p_1^2+b_{12}p_1\psi^{-1}\hat{p}_2\psi)+\frac{1}{2}b_{12}p_1\psi^{-1}\hat{p}_2\psi$$

$$+\frac{1}{2}(b_{12}\hat{p}_2p_1+b_{22}\psi^{-1}\hat{p}_2^2\psi)+V \tag{6.5}$$

$$i\hbar\frac{\partial\psi}{\partial t}=\frac{1}{2}b_{11}p_1^2\psi+\frac{1}{2}b_{12}(p_1\hat{p}_2+\hat{p}_2p_1)\psi+\frac{1}{2}b_{22}\hat{p}_2^2\psi+V\psi. \tag{6.6}$$

It will be noticed that p_1p_2 in (6.1) has been replaced by the symmetric form $1/2(p_1\hat{p}_2 + \hat{p}_2p_1)\psi$.

References

1. Wolfgang Yourgrau and Stanley Mandelstam (1968). *Variational principles in dynamics and quantum theory* (3rd edition), pp. 125–6, Pitman.
2. Cornelius Lanczos (1970). *The variational principles of mechanics*, pp. 162–3, University of Toronto Press.

Index